CONSCIOUSNESS:
A Mathematical Treatment of the
Global Neuronal Workspace Model

CONSCIOUSNESS:
A Mathematical Treatment of the
Global Neuronal Workspace Model

Rodrick Wallace, Ph.D.
Epidemiology of Mental Disorders Research Department
New York Psychiatric Institute

 Springer

Library of Congress Cataloging-in-Publication Data

A C.I.P. Catalogue record for this book is available
from the Library of Congress.

ISBN 978-1-4899-9562-9   ISBN 0-387-25244-4 (eBook)   Printed on acid-free paper.

9 8 7 6 5 4 3 2 1   SPIN 11375548

springeronline.com

BLACK NARCISSUS

Of all night's strange inhabitants,
The creature I fear worst
Never betrays the countenance
That makes my sleep accursed.

I flee and search, finding no place
His dark shape will not find,
Who lives in my own body's space
And borrows my own mind.

Alfonz Wallace

Contents

Preface

This book is not an intellectual history or popular summary of recent work on consciousness in humans. Bernard Baars (1988), Edelman and Tononi (2000), and many others, have written such, and done it well indeed. This book, rather, brings the powerful analytic machinery of communication theory to bear on the Global Neuronal Workspace (GNW) model of consciousness which Baars introduced, and does so in a formal mathematical manner.

It is not the first such attempt. The philospher Fred Dretske (1981), independent of Baars, long ago outlined how information theory might illuminate the understanding of mind. Adapting his approach on the necessary conditions for mental process, we apply a previously-developed information theory analysis of interacting cognitive biological and social modules to Baars' GNW, which has become the principal candidate for a 'standard model' of consciousness.

Invoking an obvious canonical homology with statistical physics, the method, when iterated in the spirit of the Hierarchical Linear Model of regression theory, generates a fluctuating dynamic threshold for consciousness which is similar to a tunable phase transition in a physical system. The phenomenon is, however, constrained to a manifold/atlas structure analogous to a retina; an adaptable Rate Distortion manifold, whose 'topology', in a large sense, reflects the hierarchy of embedding constraints acting on consciousness. This view greatly extends what Baars has characterized as 'contexts.'

In sharp contrast with current neural network models, our 'General Cognitive Model', and the tunable hierarchical extension which we see as the central mechanism for consciousness, can be expanded in a straightforward manner to include the influence of cognitive physiological modules like the immune system, structured psychosocial stress, or the human epigenetic system of cultural inheritance. These constraints act at a slower rate than neuronal function itself, and their inclusion produces an empirically-testable 'biopsychosociocultural' model of consciousness that meets compelling objections raised within philosophy and cultural psychology to current brain-only or hyperindividualized treatments.

The analysis can be applied to quantum systems via the quantum version of the Shannon-McMillan Theorem. Contrary to recent speculations in the physics literature, consciousness-as-we-know-it appears to be a purely classical phenomenon, with typical quantum coherence times many orders of magnitude less than the half-second which characterizes conscious behaviors. Indeed, our work suggests that the consciousness possible to a low-temperature quantum neural network would be to our own consciousness what a flask of superfluid helium is to a glass of water or a hydrogen atom to a planetary system: strange landscape indeed.

Much of the formal development has been published piecemeal elsewhere. The basic application of statistical techniques to information sources appeared in Wallace and Wallace (1998, 1999). There, renormalization at phase transition, and generalized Onsager relations away from it, were first used in the context of evolutionary theory and a model of biocultural evolution done without 'replicators.' The fundamental recognition that a class of cognitive processes has a dual information source appeared in Wallace (2000), and application to immune cognition was first made in Wallace and Wallace (2002) and Wallace (2002a). Iteration of the basic model was done in Wallace et al. (2003), examining a tunable mutator for cancer affected by psychosocial stress, and continued in Wallace and Wallace (2004) in terms of a contextually-tuned coevolutionary conflict between immune system and adaptive pathogen.

Here these developments are synthesized, extending the arguments of Wallace (2000) to second order in cognition, via punctuated universality class tuning. This produces a manifold structure whose large-scale topology is constrained by a hierarchy of contexts beginning within an individual's memory and expanding outward to include the influence of the epigenetic system of cultural heritage which so fully and completely determines the course of individual human life (Richerson and Boyd, 2004).

Dretske's necessary-condition technique is far more powerful than it may at first seem. A crude analogy is as follows: We are interested in the exact value some quantity a – analogous to consciousness – for which we have only the very limited information that $0 \leq a$. Suppose, however, we are able to determine that there exists some quantity $b \geq 0$ – analogous to the limit theorems of information theory – which we understand quite well and can indeed calculate, and that $a \leq b$. If we can know enough about b to conclude that $b \to 0$ then, of necessity, $a \equiv 0$, without requiring detailed knowledge about the behavior of a or its exact calculation: the fly-swatter argument.

As Dretske clearly saw, even 'semantically meaningful' phenomena are constrained by the asymptotic limit theorems of information theory. Here we show in some detail how this is all that is really needed to develop a profound understanding of cognitive process in general and consciousness in particular, provided one is willing to expand the perspective beyond the isolated brain.

The recommended scientific background for this book is some familiarity with the GNW model and its historical context at the level of Bernard Baars' *A Cognitive Theory of Consciousness*, as well as with current debate, for example the special issue of *Cognition* cited in the opening paragraphs of Chapter 1. The mathematical development is largely self-contained, and, for the most part, at the advanced undergraduate or beginning graduate level. Chapter 2 provides a fairly comprehensive introduction to information theory and related matters, particularly if supplemented by the standard textbooks. Chapter 4 introduces renormalization techniques in detail, and contains example calculations which should indicate ways to actually analyze empirical data, the necessary – indeed, the more important – partner to our model-driven theoretical speculations. The section of chapter 5 exploring consciousness in quantum systems can be omitted without loss of continuity.

Acknowledgments

A number of people contributed useful comments, suggestions, criticisms, and discussion, to the writing of this book. Many thanks to: B. Baars, W. Banks, P. Cohen, S. Dehaene, A. Dimitrov, M. Fullilove, J. Glazebrook, S. Heine, E. Struening, D. Wallace, and R.G. Wallace.

Chapter 1

WHAT IS CONSCIOUSNESS?

A recent, and quite remarkable, special issue of the academic journal *Cognition* (**79**(1-2), 2001)) summarizes much contemporary Western scientific work on consciousness in humans, discussing in particular a principal candidate for a new 'standard model' of the phenomenon which has been synthesized over the last two decades: the global neuronal workspace. Sergeant and Dehaene (2004) describe some of the implicit controversy as follows:

> [A growing body of empirical study shows] large all-or-none changes in neural activity when a stimulus fails to be [consciously] reported as compared to when it is reported... [A] qualitative difference between unconscious and conscious processing is generally expected by theories that view recurrent interactions between distant brain areas as a necessary condition for conscious perception... One of these theories has proposed that consciousness is associated with the interconnection of multiple areas processing a stimulus by a [dynamic] 'neuronal workspace' within which recurrent connections allow long-distance communication and auto-amplification of the activation. Neuronal network simulations... suggest the existence of a fluctuating dynamic threshold. If the primary activation evoked by a stimulus exceeds this threshold, reverberation takes place and stimulus information gains access, through the workspace, to a broad range of [other brain] areas allowing, among other processes, verbal report, voluntary manipulation, voluntary action and long-term memorization. Below this threshold, however, stimulus information remains unavailable to these processes. Thus the global neuronal workspace theory predicts an all-or-nothing transition between conscious and unconscious perception... More generally, many non-linear dynamical systems with self-amplification are characterized by the presence of discontinuous transitions in internal state...

A similar review by Baars (2002) provides a slightly different perspective, examining his own pioneering studies (Baars, 1983, 1988), along with the work of Damasio (1989), Edelman (1989), Edelman and Tononi (2000), Freeman (1991), Llinas et al. (1998), Tononi and Edelman (1998), and so on.

Baars and Franklin (2003) characterize the overall model as follows:

(1) The brain can be viewed as a collection of distributed specialized networks (processors).

(2) Consciousness is associated with a global workspace in the brain – a fleeting memory capacity whose focal contents are widely distributed ('broadcast') to many unconscious specialized networks.

(3) Conversely, a global workspace can also serve to integrate many competing and cooperating input networks.

(4) Some unconscious networks, called contexts, shape conscious contents, for example unconscious parietal maps modulate visual feature cells that underlie the perception of color in the ventral stream.

(5) Such contexts work together jointly to constrain conscious events.

(6) Motives and emotions can be viewed as goal contexts.

(7) Executive functions work as hierarchies of goal contexts.

Our particular extension of this perspective will be to introduce the idea of a hierarchical structure for 'context-of-contexts', and to explore some implications of such an extension. We thus attempt to explicitly model the roles of culture, individual developmental and community history, and embedding social network, in creating a further – and very powerful – hierarchy of constraints to conscious events.

An already large and rapidly growing body of research on the interaction of culture and individual psychology (e.g. as summarized by Markus and Kitayama, 1991; Nisbett and Masuda, 2003; Heine, 2001), which has been academically codified under the rubric of 'cultural psychology', suggests the urgent necessity of such extension.

As Heine (2001) puts the matter,

> ...[C]ultural psychology views the person as containing a set of biological potentials interacting with particular situational contexts that constrain and afford the expression of various constellations of traits and patterns of behavior. Unlike much of personality psychology, however, cultural psychology focuses on the constraints and affordances inherent to the cultural environment that give shape to those biological potentials... human nature is seen as emerging from participation in cultural worlds, and of adapting oneself to the imperatives of cultural directives...[meaning] that our nature is ultimately that of a cultural being...
>
> Cultural psychology does not view culture as a superficial wrapping of the self, or as a framework within which selves interact, but as something that is intrinsic to the self. It assumes that without culture there is no self, only a biological entity deprived of its potential... Individual selves are inextricably grounded in a configuration of consensual understandings and behavioral customs particular to a given cultural and historical context. Hence, understanding the self requires an understanding of the culture that sustains it...
>
> Cultural psychology maintains that the process of becoming a self is contingent on individuals interacting with and seizing meanings from the cultural environment...

Somewhat chillingly, Heine (2001) asserts

> The extreme nature of American individualism suggests that a psychology based on late 20th century American research not only stands the risk of developing models that are particular to that culture, but of developing an understanding of the self that is peculiar in the context of the world's cultures...

As Norenzayan and Heine (2004) point out, for the better part of a hundred years, a considerable controversy has raged within anthropology regarding the degree to which psychological and other human universals do, in fact, actually exist independent of the particularities of culture (e.g. Benedict, 1934; Mead, 1975; Geertz, 1973).

Formal application of such perspectives to theories of consciousness requires bringing together two other related strains of research on mental function and cognition taken, respectively, from physics and philosophy.

Global neuronal workspace theory has a roughly corresponding track within the physics literature, involving adaptation of a highly mathematical statistical mechanics formalism to explore observed phase transition-like behavior in the brain. These efforts range from 'bottom up' treatments by Ingber (1982, 1992) based on interacting neural network models, to the recent 'top down' mean-field approach of Steyn-Ross et al. (2001, 2003) which seeks to explain empirically observed all-or-nothing effects in general anesthesia.

Parallel to both the neuroscience and physics lines of research, but absent invocation of either dynamic systems theory or statistical mechanics, is what Adams (2003) has characterized as 'the informational turn in philosophy', that is, the application of communication theory formalism and concepts to "purposive behavior, learning, pattern recognition, and... the naturalization of mind and meaning." One of the first comprehensive attempts was that of Dretske (1981, 1988, 1992, 1993, 1994), whose work Adams describes as follows:

> It is not uncommon to think that information is a commodity generated by things with minds. Let's say that a naturalized account puts matters the other way around, viz. it says that minds are things that come into being by purely natural causal means of exploiting the information in their environments. This is the approach of Dretske as he tried consciously to unite the cognitive sciences around the well-understood mathematical theory of communication...

Dretske himself (1994) writes:

> Communication theory can be interpreted as telling one something important about the conditions that are needed for the transmission of information as ordinarily understood, about what it takes for the transmission of semantic information. This has tempted people... to exploit [information theory] in semantic and cognitive studies, and thus in the philosophy of mind.
> ...Unless there is a statistically reliable channel of communication between [a source and a receiver]... no signal can carry semantic information... [thus] the channel over which the [semantic] signal arrives [must satisfy] the appropriate statistical constraints of communication theory.

Here we redirect attention from the informational content or meaning of individual symbols, i.e. the province of semantics which so concerned Dretske, back to the statistical properties of long, internally-structured paths of symbols emitted by an information source which is 'dual' to a cognitive process in a particular sense. Application of a variety of tools adapted from statistical physics produces a dynamically tunable punctuated or phase transition coupling between interacting cognitive modules in a highly natural manner. As Dretske so clearly saw, this approach allows scientific inference on the necessary conditions for cognition, greatly illuminating the global neuronal workspace model of consciousness without raising the 18th Century ghosts of noisy, distorted mechanical clocks inherent to dynamic systems theory. The technique permits extension far beyond what is possible from statistical mechanics treatments of neural networks. In essence the method broadly recapitulates the General Linear Model (GLM) for independent or simple serially correlated observations, but on the punctuated output of an information source, using the Shannon-McMillan Theorem rather than the Central Limit Theorem. Punctuation becomes the phenomenon of central interest, rather than linear (or time series) parameter estimation.

Just as it proves fruitful to iterate the GLM, forming the Hierarchical Linear Model (HLM) in regression theory, it is possible to iterate the argument on the essential parameters of a cognitive system driving punctuation to produce the tunable, dynamic threshold so characteristic of consciousness.

The technique opens the way for the global neuronal workspace to incorporate the effects of other cognitive modules, for example the immune system, and embedding, highly structured, social or cultural contexts that may, although acting at slower timescales, greatly affect individual consciousness. These contexts-of-context function in realms beyond the brain-limited concept defined by Baars and Franklin (2003). Such extension meets profound objections to brain-only models; for example the accusation of the 'mereological fallacy' by Bennett and Hacker (2003), which we will consider in more detail later.

Before entering the formal thicket, it is important to highlight several points.

First, information theory is notorious for providing existence theorems whose representation, to use physics jargon, is arduous. For example, although the Shannon Coding Theorem implied the possibility of highly efficient coding schemes as early as 1949, it took more than forty years for practical 'turbo codes' to actually be constructed. The research program we implicitly propose here is unlikely to be any less difficult.

Second, the analysis invokes information theory variants of the fundamental limit theorems of probability. These are independent of exact mechanisms, but constrain the behavior of those mechanisms. For example, although not all processes involve long sums of independent stochastic variables, those that do, regardless of the individual variable distribution, collectively follow a Normal distribution as a consequence of the Central Limit Theorem. This has fundamental importance for estimating functional models describing relations between independent or simple serially correlated data sets – the General Linear Model and its stationary time series variants. Similarly, the games of chance in a Las Vegas casino are all quite different, but nonetheless the possible success of strategies for playing them is strongly and systematically constrained by the Martingale Theorem, regardless of game details. Analogously, languages-on-networks and languages-that-interact, as a consequence of the limit theorems of information theory, will inherently be subject to characteristic dynamic regularities regardless of detailed mechanisms, as important as the latter may be.

Just as parametric statistics are imposed, at least as a first approximation, on sometimes questionable experimental situations, relying on the robustness of the Central Limit Theorem to carry through, we take a similar heuristic approach here.

Third, this work invokes an obvious homology between information source uncertainty and thermodynamic free energy density as justification for importing renormalization and generalized Onsager relation formalism to the study of cognitive process near and away from 'critical points' in the coupling of cognitive submodules. The purpose is to create a 'General Cognitive Model' (GCM) for the punctuated behavior of cognitive phenomena constrained by the Shannon-McMillan Theorem, a model which permits estimation of essential system parameters from observational data.

The question of whether the analysis demonstrates the necessity of global phase transitions for information-transmission networks or merely builds a suggestive analogy with thermodynamics is, of course, ultimately empirical. Can the model actually be used to analyze experimental or observational data? For the microscopic case, however, Feynman (1996) has shown that the homology we invoke is an identity, which is no small matter and indeed suggests that behavior analogous to phase transitions in simple physical systems should be ubiquitous for a very broad class of information systems.

Our work appears similar, in a certain sense, to Bohr's treatment of the atom, which attempted a simple substitution of quantized angular momentum into a basically classical theory. Although incomplete, his analysis contributed materially to the more comprehensive approach of quantum mechanics. In that spirit, increasingly satisfactory models may follow from the interplay of our work here and appropriate empirical studies.

In just this regard, it is worth reiterating the dire warnings of the mathematical ecologist E.C. Pielou on the place of mathematical models in studies of complicated systems (Pielou, 1977, p. 106):

> ...[Mathematical] models are easy to devise; even though the assumptions of which they are constructed may be hard to justify, the magic phrase 'let us assume that...' overrides objections temporarily. One is then confronted with a much harder task: How is such a model to be tested? The correspondence between a model's predictions and

observed events is sometimes gratifyingly close but this cannot be taken to imply the
model's simplifying assumptions are reasonable in the sense that neglected complications are indeed negligible in their effects...

In my opinion the usefulness of models is great... [however] it consists *not in answer-
ing questions but in raising them.* Models can be used to inspire new field investigations
and these are the only source of new knowledge as opposed to new speculation.

We will use a mathematical model of the global neuronal workspace to
speculate in some detail on the relations between consciousness and social
and cultural contexts, speculation which hopefully will inspire new empirical
investigations and, through these, a new round of interaction between models
and experiments.

The next chapter recapitulates some facts and theorems from information
theory and related disciplines. The third gets down to the serious business of
describing cognitive process in terms of an information source, a kind of language constrained by the Shannon-McMillan Theorem, and its Rate Distortion
or Joint Asymptotic Equipartition and other variants for interacting sources.
Subsequent chapters introduce phase transition formalism, and apply it to the
effects of social and cultural contexts on individual consciousness.

Chapter 2

INFORMATION THEORY

Suppose that the probability of interaction between individuals (or extended families or other organizational nodes) linked in a network depends jointly on their *geographic* and *social* locations, which we characterize as multidimensional vector quantities $\mathbf{X}$ and $\mathbf{Z}$ respectively. These measures might be determined, for example, from survey data on individuals or inferred from environmental index data on groups. Thus for individual nodes j and k we assume their probability of interaction $P_{j,k}$ is given by

$$P_{j,k} = P_{j,k}(\mathbf{X}_j, \mathbf{X}_k, \mathbf{Z}_j, \mathbf{Z}_k)$$

where $0 \leq P_{j,k} \leq 1$.

It may be possible to reduce $P_{j,k}$ to a function of the differences $\mathbf{X} = \mathbf{X}_j - \mathbf{X}_k$ and $\mathbf{Z} = \mathbf{Z}_j - \mathbf{Z}_k$, or, perhaps, by using a multivariate method such as principal components analysis, even to functions of their 'length' $x = |\mathbf{X}|$ and $z = |\mathbf{Z}|$, so that

$$P_{j,k} = P_{j,k}(x, z).$$

One way to proceed is to impose a generalized distance $r^2 \equiv x^2 + z^2$, and explore the effects of various probability distributions which are functions of r. This is, in fact, done at some length in Chapter 4. Here, rather, we finesse the argument and transform out of the space defined by x and z into the probability space itself, defining a metric according to

$$L_{j,k} \equiv \log(1/P_{j,k})$$

(2.1)

where log is the logarithm to some base number. Note that it is the probability distribution based on the generalized distance r which induces the transformation between 'real' space and probability space.

If it can be assumed for all nodes j, k, l within a sufficiently small network patch, that

$$P_{j,l} \geq P_{j,k} P_{k,l},$$

so that

$$\frac{1}{P_{j,l}} \leq \frac{1}{P_{j,k}} \frac{1}{P_{k,l}}$$

and the strong 'triangle' inequality

$$\log(1/P_{j,l}) \leq \log(1/P_{j,k}) + \log(1/P_{k,l})$$

holds, then L is a pseudometric, and various standard attacks are possible.

That somewhat draconian condition can, however, be considerably weakened as follows:

Let ΔL_j be the average 'distance' in probability space from the node j to all other nodes, that is,

$$\Delta L_j \equiv \sum_k P_{j,k} \log(1/P_{j,k})$$

(2.2)

Suppose some fairly elaborate 'message,' not otherwise characterized, is sent along the sociogeographic network, and a *traveling wave* condition is imposed, so that, for some time period Δt, the relation

$$\frac{\Delta L_j}{\Delta t} \approx C,$$

(2.3)

holds. C is, then, *the mean fixed rate at which the message is sent from the network to an embedded individual.*

Wallace et al, (1996) show – not unexpectedly and probably not originally – that the traveling wave assumption, on a fractal manifold, is equivalent to the Aharony-Stauffer conjecture, which directly relates the fractal dimension of the interior of an affected network to that of its growth surface. This gives a simple and explicit expression for C, usually an arduous calculation. The detailed derivation is left as an exercise.

By expanding equation (2.3) we obtain, taking $\Delta t \equiv 1$,

$$- \sum_k P_{j,k} \log(P_{j,k}) \approx C$$

(2.4)

where C is a transmission rate constant characteristic of the particular sociogeographic network. We further assume that $\sum_k P_{j,k} \equiv 1$, i.e., the network is 'tight' in the sense that each node interacts with it as a whole with unit probability. Hence $P_{j,k}$ is a legitimate probability distribution.

These are deep waters: For any probability distribution, $0 \leq P_j \leq 1$ such that $\sum_j P_j = 1$ the quantity

$$H = - \sum_j P_j \log(P_j)$$

(2.5)

is the distribution's *Shannon uncertainty*, a fundamental quantity of classical information theory.

Neglecting details explored below, the transfer of uncertainty represents the transmission of information: The Shannon Coding Theorem, the first important result of information theory, states that for any rate $R < C$, where C represents the capacity of the information channel, it is possible to find a 'coding scheme' such that a sufficiently long message can be sent with arbitrarily small error.

This is surely one of the most striking conclusions of 20th Century applied mathematics.

1.   The Shannon Coding Theorem

Messages from a source, seen as symbols x_j from some alphabet, each having probabilities P_j associated with a random variable X, are 'encoded' into the language of a 'transmission channel', a random variable Y with symbols y_k, having probabilities P_k, possibly with error. Someone receiving the symbol y_k then retranslates it (without error) into some x_k, which may or may not be the same as the x_j that was sent.

More formally, the message sent along the channel is characterized by a random variable X having the distribution

$$P(X = x_j) = P_j, j = 1, ..., M.$$

The channel through which the message is sent is characterized by a second random variable Y having the distribution

$$P(Y = y_k) = P_k, k = 1, ..., L.$$

Let the joint probability distribution of X and Y be defined as

$$P(X = x_j, Y = y_k) = P(x_j, y_k) = P_{j,k}$$

and the conditional probability of Y given X as

$$P(Y = y_k | X = x_j) = P(y_k | x_j).$$

Then the Shannon uncertainty of X and Y independently and the joint uncertainty of X and Y together are defined respectively as

$$H(X) = -\sum_{j=1}^{M} P_j \log(P_j)$$

$$H(Y) = -\sum_{k=1}^{L} P_k \log(P_k)$$

$$H(X,Y) = -\sum_{j=1}^{M} \sum_{k=1}^{L} P_{j,k} \log(P_{j,k}).$$

(2.6)

The *conditional uncertainty* of Y given X is defined as

$$H(Y|X) = -\sum_{j=1}^{M}\sum_{k=1}^{L} P_{j,k} \log[P(y_k|x_j)]$$

(2.7)

For any two stochastic variates X and Y, $H(Y) \geq H(Y|X)$, as knowledge of X generally gives some knowledge of Y. Equality occurs only in the case of stochastic independence.

Since $P(x_j, y_k) = P(x_j)P(y_k|x_j)$, we have

$$H(X|Y) = H(X,Y) - H(Y)$$

The information transmitted by translating the variable X into the channel transmission variable Y – possibly with error – and then retranslating without error the transmitted Y back into X is defined as

$$I(X|Y) \equiv H(X) - H(X|Y) = H(X) + H(Y) - H(X,Y)$$

(2.8)

See, for example, Ash (1990), Khinchine (1957) or Cover and Thomas (1991) for details. The essential point is that if there is no uncertainty in X given the channel Y, then there is no loss of information through transmission.

In general this will not be true, and herein lies the essence of the theory.

Given a fixed vocabulary for the transmitted variable X, and a fixed vocabulary and probability distribution for the channel Y, we may vary the probability distribution of X in such a way as to maximize the information sent. The capacity of the channel is defined as

$$C \equiv \max_{P(X)} I(X|Y)$$

(2.9)

subject to the subsidiary condition that $\sum P(X) = 1$.

The critical trick of the Shannon Coding Theorem for sending a message with arbitrarily small error along the channel Y at any rate $R < C$ is to encode it in longer and longer 'typical' sequences of the variable X; that is, those sequences whose distribution of symbols approximates the probability distribution $P(X)$ above which maximizes C.

If $S(n)$ is the number of such 'typical' sequences of length n, then

$$\log[S(n)] \approx nH(X)$$

where $H(X)$ is the uncertainty of the stochastic variable defined above. Some consideration shows that $S(n)$ is much less than the total number of possible messages of length n. Thus, as $n \to \infty$, only a vanishingly small fraction of all possible messages is meaningful in this sense. This observation, after some considerable development, is what allows the Coding Theorem to work so well. In sum, the prescription is to encode messages in typical sequences, which are sent at very nearly the capacity of the channel. As the encoded messages become longer and longer, their maximum possible rate of transmission without error approaches channel capacity as a limit. Again, Ash (1990), Khinchine (1957) and Cover and Thomas (1991) provide details.

2.   More heuristics: a 'tuning theorem'

Telephone lines, optical wave guides and the tenuous plasma through which a planetary probe transmits data to earth may all be viewed in traditional information-theoretic terms as a *noisy channel* around which we must structure a message so as to attain an optimal error-free transmission rate.

Telephone lines, wave guides and interplanetary plasmas are, relatively speaking, fixed on the timescale of most messages, as are most sociogeographic networks. Indeed, the capacity of a channel, according to equation (2.9), is defined by varying the probability distribution of the 'message' process X so as to maximize $I(X|Y)$.

Suppose there is some message X so critical that its probability distribution must remain fixed. The trick is to fix the distribution $P(x)$ but *modify the channel* – i.e. tune it – so as to maximize $I(X|Y)$. The *dual* channel capacity C^* can be defined as

$$C^* \equiv \max_{P(Y),P(Y|X)} I(X|Y)$$

(2.10)

But

$$C^* = \max_{P(Y),P(Y|X)} I(Y|X)$$

since

$$I(X|Y) = H(X) + H(Y) - H(X,Y) = I(Y|X).$$

Thus, in a purely formal mathematical sense, *the message transmits the channel*, and there will indeed be, according to the Coding Theorem, a channel distribution $P(Y)$ which maximizes C^*.

One may do better than this, however, by modifying the channel matrix $P(Y|X)$. Since

$$P(y_j) = \sum_{i=1}^{M} P(x_i)P(y_j|x_i),$$

$P(Y)$ is entirely defined by the channel matrix $P(Y|X)$ for fixed $P(X)$ and

$$C^* = \max_{P(Y),P(Y|X)} I(Y|X) = \max_{P(Y|X)} I(Y|X).$$

Calculating C^* requires maximizing the complicated expression

$$I(X|Y) = H(X) + H(Y) - H(X,Y)$$

which contains products of terms and their logs, subject to constraints that the sums of probabilities are 1 and each probability is itself between 0 and 1. Maximization is done by varying the channel matrix terms $P(y_j|x_i)$ within the constraints. This is a difficult problem in nonlinear optimization. See Parker et al. (2003) for a comprehensive treatment, using traditional Lagrange multiplier methods. However, for the special case $M = L$, C^* may be found by inspection:

If $M = L$, then choose

$$P(y_j|x_i) = \delta_{j,i}$$

where $\delta_{i,j}$ is 1 if $i = j$ and 0 otherwise. For this special case

$$C^* \equiv H(X)$$

with $P(y_k) = P(x_k)$ for all k. *Information is thus transmitted without error when the channel becomes 'typical' with respect to the fixed message distribution $P(X)$.*

If $M < L$ matters reduce to this case, but for $L < M$ information must be lost, leading to 'Rate Distortion' arguments explored more fully below.

Thus modifying the channel may be a far more efficient means of ensuring transmission of an important message than encoding that message in a 'natural' language which maximizes the rate of transmission of information on a fixed channel.

We have examined the two limits in which either the distributions of $P(Y)$ or of $P(X)$ are kept fixed. The first provides the usual Shannon Coding Theorem, and the second, hopefully, a tuning theorem variant. It seems likely, however, than for many important systems $P(X)$ and $P(Y)$ will 'interpenetrate,' to use Richard Levins' terminology. That is, $P(X)$ and $P(Y)$ will affect each other in characteristic ways, so that some form of mutual tuning may be the most effective strategy.

3.   The Shannon-McMillan Theorem

Not all statements – sequences of the random variable X – are equivalent. According to the structure of the underlying language of which the message is a particular expression, some messages are more 'meaningful' than others, that is, in accord with the grammar and syntax of the language. The other principal result from information theory, the Shannon-McMillan or Asymptotic Equipartition Theorem, describes how messages themselves are to be classified.

Suppose a long sequence of symbols is chosen, using the output of the random variable X above, so that an output sequence of length n, with the form

$$x_n = (\alpha_0, \alpha_1, ..., \alpha_{n-1})$$

has joint and conditional probabilities

$$P(X_0 = \alpha_0, X_1 = \alpha_1, ..., X_{n-1} = \alpha_{n-1})$$

$$P(X_n = \alpha_n | X_0 = \alpha_0, ..., X_{n-1} = \alpha_{n-1}).$$

(2.11)

Using these probabilities we may calculate the conditional uncertainty

$$H(X_n|X_0, X_1, ..., X_{n-1}).$$

The uncertainty of the *information source*, $H[\mathbf{X}]$, is defined as

$$H[\mathbf{X}] \equiv \lim_{n \to \infty} H(X_n|X_0, X_1, ..., X_{n-1}).$$

(2.12)

In general

$$H(X_n|X_0, X_1, ..., X_{n-1}) \leq H(X_n).$$

Only if the random variables X_j are all stochastically independent does equality hold. If there is a maximum n such that, for all $m > 0$

$$H(X_{n+m}|X_0, ..., X_{n+m-1}) = H(X_n|X_0, ..., X_{n-1}),$$

then the source is said to be of *order* n. It is easy to show that

$$H[\mathbf{X}] = \lim_{n \to \infty} \frac{H(X_0, ...X_n)}{n+1}.$$

In general the outputs of the $X_j, j = 0, 1, ..., n$ are *dependent*. That is, the output of the communication process at step n depends on previous steps. Such serial correlation, in fact, is the very structure which enables most of what follows in this book.

Here, however, the processes are all assumed stationary in time, that is, the serial correlations do not change in time, and the system is *memoryless*.

A very broad class of such self-correlated, memoryless, information sources, the so-called *ergodic* sources for which the long-run relative frequency of a sequence converges stochastically to the probability assigned to it, have a particularly interesting property:

It is possible, in the limit of large n, to divide all sequences of outputs of an ergodic information source into two distinct sets, S_1 and S_2, having, respectively, very high and very low probabilities of occurrence, with the source uncertainty providing the splitting criterion. In particular the Shannon-McMillan Theorem states that, for a (long) sequence having n (serially correlated) elements, the number of 'meaningful' sequences, $N(n)$ – those belonging to set S_1 – will satisfy the relation

$$\frac{\log[N(n)]}{n} \approx H[\mathbf{X}].$$

(2.13)

More formally,

$$\lim_{n \to \infty} \frac{\log[N(n)]}{n} = H[\mathbf{X}]$$

$$= \lim_{n \to \infty} H(X_n | X_0, ..., X_{n-1})$$

$$= \lim_{n \to \infty} \frac{H(X_0, ..., X_n)}{n+1}.$$

(2.14)

The Shannon Coding theorem, by means of an analogous splitting argument, shows that for any rate $R < C$, where C is the channel capacity, a message may be sent without error, using the probability distribution for X which maximizes $I(X|Y)$ as the coding scheme. Using the internal structures of the information source permits *limiting attention only to meaningful sequences of symbols*. This restriction can greatly raise the maximum possible rate at which information can be transmitted with arbitrarily small error: if there are M possible symbols and the uncertainty of the source is $H[\mathbf{X}]$, then the effective capacity of the channel C, using this 'source coding,' becomes (Ash, 1990)

$$C_E = C \frac{\log(M)}{H[\mathbf{X}]}.$$

(2.15)

As $H[\mathbf{X}] \leq \log(M)$, with equality only for stochastically independent, uniformly distributed random variables,

$$C_E \geq C.$$

(2.16)

Note that, for a given channel capacity, the condition

$$H[\mathbf{X}] \leq C$$

always holds.

Source uncertainty has a very important heuristic interpretation. As Ash (1990) puts it,

> ...[W]e may regard a portion of text in a particular language as being produced by an information source. The probabilities $P[X_n = \alpha_n | X_0 = \alpha_0, ..., X_{n-1} = \alpha_{n-1})$ may be estimated from the available data about the language; in this way we can estimate the uncertainty associated with the language. A large uncertainty means, by the [Shannon-McMillan Theorem], a large number of 'meaningful' sequences. Thus given two languages with uncertainties H_1 and H_2 respectively, if $H_1 > H_2$, then in the absence of noise it is easier to communicate in the first language; more can be said in the same amount of time. On the other hand, it will be easier to reconstruct a scrambled portion of text in the second language, since fewer of the possible sequences of length n are meaningful.

It is possible to significantly generalize this heuristic picture in such a way as to characterize the interaction between different 'languages,' something at the core of the development.

4. The Rate Distortion Theorem

The Shannon-McMillan Theorem can be expressed as the 'zero error limit' of something called the Rate Distortion Theorem (Dembo and Zeitouni, 1998; Cover and Thomas, 1991), which defines a splitting criterion that identifies high probability pairs of sequences. We follow closely the treatment of Cover and Thomas (1991).

The origin of the problem is the question of representing one information source by a simpler one in such a way that the least information is lost. For example we might have a continuous variate between 0 and 100, and wish to represent it in terms of a small set of integers in a way that minimizes the inevitable distortion that process creates. Typically, for example, an analog audio signal will be replaced by a 'digital' one. The problem is to do this in a way which least distorts the *reconstructed* audio waveform.

Suppose the original memoryless, ergodic information source Y with output from a particular alphabet generates sequences of the form

$$y^n = y_1, ..., y_n.$$

These are 'digitized,' in some sense, producing a chain of 'digitized values'

$$b^n = b_1, ..., b_n,$$

where the b-alphabet is much more restricted than the y-alphabet.

b^n is, in turn, *deterministically retranslated* into a reproduction of the original signal y^n. That is, each b^m is mapped on to a unique n-length y-sequence in the alphabet of the information source Y:

$$b^m \rightarrow \hat{y}^n = \hat{y}_1, ..., \hat{y}_n.$$

Note, however, that many y^n sequences may be mapped onto the *same* retranslation sequence $\hat{y}^n$, so that information will, in general, be lost.

The central problem is to explicitly minimize that loss.

The retranslation process defines a new memoryless, ergodic information source, $\hat{Y}$.

The next step is to define a *distortion measure*, $d(y, \hat{y})$, which compares the original to the retranslated path. For example the *Hamming distortion* is

$$d(y, \hat{y}) = 1, y \neq \hat{y}$$

$$d(y, \hat{y}) = 0, y = \hat{y}.$$

(2.17)

For continuous variates the *Squared error distortion* is

$$d(y, \hat{y}) = (y - \hat{y})^2.$$

(2.18)

Possibilities abound.

The distortion between paths y^n and $\hat{y}^n$ is defined as

$$d(y^n, \hat{y}^n) = \frac{1}{n} \sum_{j=1}^{n} d(y_j, \hat{y}_j).$$

(2.19)

Suppose that with each path y^n and b^n-path retranslation into the y-language and denoted y^n, there are associated individual, joint, and conditional probability distributions

$$p(y^n), p(\hat{y}^n), p(y^n | \hat{y}^n).$$

The *average distortion* is defined as

$$D = \sum_{y^n} p(y^n) d(y^n, \hat{y}^n).$$

(2.20)

It is possible, using the distributions given above, to define the information transmitted from the incoming Y to the outgoing $\hat{Y}$ process in the usual manner, using the Shannon source uncertainty of the strings:

$$I(Y, \hat{Y}) \equiv H(Y) - H(Y | \hat{Y}) = H(Y) + H(\hat{Y}) - H(Y, \hat{Y}).$$

If there is no uncertainty in Y given the retranslation $\hat{Y}$, then no information is lost.

In general, this will not be true.

The *information rate distortion function* $R(D)$ for a source Y with a distortion measure $d(y, \hat{y})$ is defined as

$$R(D) = \min_{p(y,\hat{y});\sum_{(y,\hat{y})} p(y)p(y|\hat{y})d(y,\hat{y})\leq D} I(Y,\hat{Y}).$$

(2.21)

The minimization is over all conditional distributions $p(y|\hat{y})$ for which the joint distribution $p(y, \hat{y}) = p(y)p(y|\hat{y})$ satisfies the average distortion constraint (i.e. average distortion $\leq D$).

The *Rate Distortion Theorem* states that $R(D)$ *is the maximum achievable rate of information transmission which does not exceed the distortion D*. Cover and Thomas (1991) or Dembo and Zeitouni (1998) provide details, and Parker et al. (2003) formalize a comprehensive attack.

More to the point, however, is the following: Pairs of sequences $(y^n, \hat{y}^n)$ can be defined as *distortion typical*; that is, for a given average distortion D, defined in terms of a particular measure, pairs of sequences can be divided into two sets, a high probability one containing a relatively small number of (matched) pairs with $d(y^n, \hat{y}^n) \leq D$, and a low probability one containing most pairs. As $n \to \infty$, the smaller set approaches unit probability, and, for those pairs,

$$p(y^n) \geq p(\hat{y}^n|y^n) \exp[-nI(Y,\hat{Y})].$$

(2.22)

Thus, roughly speaking, $I(Y,\hat{Y})$ embodies the splitting criterion between high and low probability pairs of paths.

For the theory of interacting information sources, then, $I(Y,\hat{Y})$ can play the role of H in the dynamic treatment that follows.

The rate distortion function of eq. 2.21 can actually be calculated in many cases by using a Lagrange multiplier method – see Section 13.7 of Cover and Thomas (1991).

At various points in the development we will suggest using $s \equiv d(\hat{x}, x)$ as a metric in a geometry of information sources, e.g. when simple ergodicity fails, and $H(x) \neq H(\hat{x})$ for high probability paths $\hat{x}$ and x. See eq. (3.2).

5.   Large Deviations

The use of information source uncertainty above as a splitting criterion between high and low probability sequences (or pairs of them) displays the fundamental characteristic of a growing body of work in applied probability often termed the 'Large Deviations Program,' (LDP) which seeks to unite information theory, statistical mechanics and the theory of fluctuations under a single umbrella. It serves as a convenient starting point for further developments.

We can begin to place information theory in the context of the LDP as follows (Dembo and Zeitouni, 1998, p.2):

Let $X_1, X_2, ...X_n$ be a sequence of independent, standard Normal, real-valued random variables and let

$$S_n = \frac{1}{n} \sum_{j=1}^{n} X_j.$$

(2.23)

Since S_n is again a Normal random variable with zero mean and variance $1/n$, for all $\delta > 0$

$$\lim_{n \to \infty} P(|S_n| \geq \delta) = 0,$$

(2.24)

where P is the probability that the absolute value of S_n is greater or equal to δ. Some manipulation, however, gives

$$P(|S_n| \geq \delta) = 1 - \frac{1}{\sqrt{2\pi}} \int_{-\delta\sqrt{n}}^{\delta\sqrt{n}} \exp(-x^2/2)dx,$$

(2.25)

so that

$$\lim_{n \to \infty} \frac{\log P(|S_n| \geq \delta)}{n} = -\delta^2/2$$

(2.26)

This can be rewritten for large n as

$$P(|S_n| \geq \delta) \approx \exp(-n\delta^2/2).$$

(2.27)

That is, for large n, the probability of a large deviation in S_n follows something much like equation (2.13), i.e. that meaningful paths of length n all have approximately the same probability $P(n) \propto \exp(-nH[\mathbf{X}])$.

Our questions about 'meaningful paths' appear suddenly as formally isomorphic to the central argument of the LDP which encompasses statistical mechanics, fluctuation theory, and information theory into a single structure (Dembo and Zeitouni, 1998).

A cardinal tenet of large deviation theory is that the 'rate function' $-\delta^2/2$ in equation (2.26) can, under proper circumstances, be expressed as a mathematical 'entropy' having the standard form

$$-\sum_k p_k \log p_k,$$

(2.28)

for some set of probabilities p_k. This striking result goes under various names at various levels of approximation – Sanov's Theorem, Cramer's Theorem, the Gartner-Ellis Theorem, the Shannon-McMillan Theorem, and so on (Dembo and Zeitouni, 1998).

6. Fluctuations

The standard treatment of 'fluctuations' (Onsager and Machlup, 1953; Fredlin and Wentzell, 1998) in physical systems is the principal foundation for much current study of stochastic resonance and related phenomena and also serves as a useful reference point.

The macroscopic behavior of a complicated physical system in time is assumed to be described by the phenomenological Onsager relations giving large-scale fluxes as

$$\sum_i R_{i,j} dK_j/dt = \partial S/\partial K_i,$$

(2.29)

where the $R_{i,j}$ are appropriate constants, S is the system entropy and the K_i are the generalized coordinates which parametize the system's free energy.

Entropy is defined from free energy F by a Legendre transform – more of which follows below:

$$S \equiv F - \sum_j K_j \partial F/\partial K_j,$$

where the K_j are appropriate system parameters.

Neglecting volume problems (which will become quite important later), free energy can be defined from the system's partition function Z as

$$F(K) = \log[Z(K)].$$

The partition function Z, in turn, is defined from the system Hamiltonian – defining the energy states – as

$$Z(K) = \sum_j \exp[-KE_j],$$

where K is an inverse temperature or other parameter and the E_j are the energy states.

Inverting the Onsager relations gives

$$dK_i/dt = \sum_j L_{i,j} \partial S/\partial K_j = L_i(K_1, ..., K_m, t) \equiv L_i(K, t).$$

(2.30)

The terms $\partial S/\partial K_i$ are macroscopic driving 'forces' dependent on the entropy gradient.

Let a white Brownian 'noise' $\epsilon(t)$ perturb the system, so that

$$dK_i/dt = \sum_j L_{i,j}\partial S/\partial K_j + \epsilon(t)$$

$$= L_i(K,t) + \epsilon(t),$$

(2.31)

where the time averages of ϵ are $< \epsilon(t) >= 0$ and $< \epsilon(t)\epsilon(0) >= D\delta(t)$. $\delta(t)$ is the Dirac delta function, and we take K as a vector in the K_i.

Following Luchinsky (1997), if the probability that the system starts at some initial macroscopic parameter state K_0 at time $t = 0$ and gets to the state $K(t)$ at time t is $P(K,t)$, then a somewhat subtle development (e.g. Feller, 1971) gives the forward Fokker-Planck equation for P:

$$\partial P(K,t)/\partial t = -\nabla \cdot (L(K,t)P(K,t)) + (D/2)\nabla^2 P(K,t).$$

(2.32)

In the limit of weak noise intensity this can be solved using the WKB, i.e. the eikonal, approximation, as follows: take

$$P(K,t) = z(K,t)\exp(-s(K,t)/D).$$

(2.33)

$z(K, t)$ is a prefactor and $s(K, t)$ is a classical action satisfying the Hamilton-Jacobi equation, which can be solved by integrating the Hamiltonian equations of motion. The equation reexpresses $P(K, t)$ in the usual parametized negative exponential format.

Let $p \equiv \nabla s$. Substituting equation (2.33) in equation (2.32) and collecting terms of similar order in D gives

$$dK/dt = p + L, dp/dt = -\partial L/\partial K p$$

$$-\partial s/\partial t \equiv h(K, p, t) = pL(K, t) + \frac{p^2}{2},$$

with $h(K, t)$ the 'Hamiltonian' for appropriate boundary conditions.

Again following Luchinsky (1997), these 'Hamiltonian' equations have two different types of solution, depending on p. For $p = 0$, $dK/dt = L(K, t)$ which describes the system in the absence of noise. We expect that with finite noise intensity the system will give rise to a distribution about this deterministic path. Solutions for which $p \neq 0$ correspond to *optimal paths* along which the system will move with overwhelming probability.

This is a formulation of fluctuation theory which has particular attraction for physicists, few of whom can resist the nearly magical appearance of a Hamiltonian. These results can, however, again be directly derived as a special case of a Large Deviation Principle based on 'generalized 'entropies' mathematically similar to Shannon's uncertainty from information theory, bypassing the 'Hamiltonian' formulation entirely (Dembo and Zeitouni, 1998).

For languages, of course, there is no possibility of a Hamiltonian, but the generalized entropy or splitting criterion treatment still works. The trick will be to do with entropies what is most often done with Hamiltonians:

Here we will be concerned, not with a random Brownian distortion of simple physical systems, but with a complex 'behavioral' structure, in the largest sense, composed of quasi-independent 'actors' for which

[1] the usual Onsager relations of equations (2.29) and (2.30) may be too simple,

[2] the 'noise' may not be either small or random, and, most critically,

[3] *the meaningful/optimal paths have extremely structured serial correlation, amounting to a grammar and syntax, precisely the fact which allows definition of an information source* and enables the use of the very sparse equipartition of the Shannon-McMillan and Rate Distortion Theorems. The sparseness and equipartition, in fact, permit solution of the problems we will address.

In sum, to again paraphrase Luchinsky (1997), large fluctuations, although infrequent, are fundamental in a broad range of processes, and it was recognized by Onsager and Machlup (1953) that insight into the problem could be gained from studying the distribution of fluctuational paths along which the system moves to a given state. This distribution is a fundamental characteristic of the fluctuational dynamics, and its understanding leads toward control of fluctuations. Fluctuational motion from the vicinity of a stable state may occur along different paths. For large fluctuations, the distribution of these paths peaks sharply along an optimal, most probable, path. In the theory of large fluctuations, the pattern of optimal paths plays a role similar to that of the phase portrait in nonlinear dynamics.

In this development 'meaningful' paths play the role of 'optimal' paths in the theory of large fluctuations, but without benefit of a 'Hamiltonian.'

7.   The fundamental homology

Section 5 above gives something of the flavor of the LDP which tries to unify statistical mechanics, large fluctuations and information theory. This opens a methodological Pandora's Box: the LDP provides justification for a massive transfer of superstructure from statistical mechanics to information theory, including real-space renormalization for address of phase transition, thermodynamics and an equation of state, generalized Onsager relations, and so on. From fluctuation theory and nonlinear dynamics come phase space, domains of attraction and related matters.

Several particulars distinguish this approach.

First is a draconian simplification which seeks to employ information theory concepts only as they directly relate to the basic limit theorems of the subject. That is, message uncertainty and information source uncertainty are interesting only because they obey the Coding, Source Coding, Rate Distortion, and related theorems. 'Information Theory' treatments which do not sufficiently center on these theorems are, from this view, far off the mark. Thus most discussion of 'complexity,' 'entropy maximization,' different definitions of 'entropy,' and so forth, just does not appear on the horizon. In the words of William of Occam, "Entities ought not be multiplied without necessity."

The second matter is somewhat more complicated: Rojdestvenski and Cottam (2000, p.44), following Wallace and Wallace (1998), see the linkage between information theory and statistical mechanics as a characteristic

> ...[homological] mapping... between... unrelated... problems that share the same mathematical basis... [whose] similarities in mathematical formalisms...become powerful tools for [solving]... traditional problems.

The possible relation of information theory to biological and social process, both of which can involve agency, appears very sharply constrained, involving:

(1) a 'linguistic' equipartition of sets of probable paths consistent with the Shannon-McMillan, Rate Distortion, or related theorems which serves as the formal connection with nonlinear mechanics and fluctuation theory, and

(2) a homological correspondence between information source uncertainty and statistical mechanical free energy density, not statistical mechanical entropy.

In this latter regard, the definition of the free energy density of a parametized physical system is

$$F(K_1, ..., K_m) = \lim_{V \to \infty} \frac{\log[Z(K_1, ..., K_m, V)]}{V},$$

(2.34)

where the K_j are parameters, V is the system volume, and Z is, again, the partition function.

For an ergodic information source the equivalent relation associates the source uncertainty with the number of 'meaningful' statements $N(n)$ of length n, in the limit,

$$H[\mathbf{X}] = \lim_{n \to \infty} \frac{\log[N(n)]}{n}.$$

This can be parametized in various manners to obtain the crucial expression on which all else is built:

$$H[K_1, ..., K_m, \mathbf{X}] = \lim_{n \to \infty} \frac{\log[N(K_1, ..., K_m, n)]}{n}.$$

(2.35)

At first glance, Shannon uncertainty has the algebraic form of the entropy of a physical system, $\propto \sum_k P_k \log(P_k)$, where the P_k constitute a probability distribution. This is deceptive. In the absence of a 'distinguishing two-form' which defines a 'symplectic geometry', that is, in the absence of a second order Hamiltonian defining energy, Shannon uncertainty cannot be the 'entropy' of a system, even if it has the same mathematical form. See Arnold (1989) for an explanation of the 'symplectic' jargon, but the basic point is that the concept of entropy is directly derived from ideas of work and heat, while Shannon uncertainty has its origin in the process of sending a message.

While it is sometimes possible to impose a distinguishing two-form on a contact manifold, to symplectify it by artificially constructing an analog to a Hamiltonian (Arnold, 1989), such logical convolutions are not really needed. In any event, when such a 'duality' is invoked, Shannon uncertainty does not become the analog of thermodynamic entropy: resolution of a famous paradox in physics requires an identification of Shannon uncertainty with free energy density. As Elitzur (1996, p. 179) puts it

> Recall ... the lesson of Maxwell's Demon: Information, when applied under the appropriate circumstances, can save work.

Bennett (1988, p. 230), as quoted by Elitzur (1996) states

> ...[T]he value of a message is the amount of mathematical or other work plausibly done by its originator, which the receiver is saved from having to repeat.

Similarly, Feynman (1996) provides a formal example, showing that, for a certain class of microscopic systems, transmission of information can be interpreted as exchange of free energy.

This, then, is the essential 'homology' linking information theory to the technology of statistical mechanics and related disciplines. Only for very simple systems – e.g. Bennett's microscopically reversible computing machinery – can the homology be an identity. In general this will not be the case, as individual 'agency' increasingly imposes behavioral regularities which are not simply mechanistic: Thus the 'Hamiltonian' goes away, but an 'entropy' treatment using source uncertainties, remains possible.

That is, for mesoscale or 'behavioral' systems, infinite-volume-based or Hamiltonian-driven thermodynamic treatments familiar from physics are inappropriate since either the usual forms of the ergodic theorem break down (e.g. Bar-Yam, 1997), or there is simply no underlying scalar function to maximize or minimize. It is possible to regain something much like the ergodic theorem for such phenomena by imposing the grammar and syntax inherent in the Shannon-McMillan or the Rate Distortion Theorems through the limit relations defining the splitting between high and low probability sequences,

$$H[\mathbf{X}] = \lim_{n \to \infty} \frac{\log[N(n)]}{n}.$$

In the context of an appropriate parametization, a kind of thermodynamic formalism can, then, also be imposed, but the results will usually have little relation to ordinary thermodynamics, particularly for the usual energetically open systems of most interest.

The next task is to reexamine cognitive process from an information theory perspective.

Chapter 3

COGNITION AS GENERALIZED LANGUAGE

1. Theory

Atlan and Cohen (1998) and Cohen (2000) argue that the essence of cognitive function involves comparison of a perceived signal with an internal, learned picture of the world, and then, upon that comparison, choice of one response from a much larger repertorie of possible responses. Their analysis is in a long tradition of speculation regarding immune cognition (e.g. Grossman, 1989; Tauber, 1998). Taking the approach of Wallace (2000, 2002a), we make a 'weak', and hence very general, model of that process, which will be illustrated by two neural network examples.

Cognitive pattern recognition-and-response, from this perspective, proceeds by convoluting an incoming external sensory signal with an internal ongoing activity – the learned picture of the world – and triggering an appropriate action based on a decision that the pattern of sensory activity requires a response. We will, fulfilling Atlan and Cohen's (1998) criterion of meaning-from-response, define a language's contextual meaning entirely in terms of system output, neglecting, for the moment, the question of how such a pattern recognition system is trained, a matter for Rate Distortion Theory.

A pattern of sensory input is, then, mixed in an unspecified but systematic manner with an internal 'ongoing' activity to create a path of convoluted signals $x = (a_0, a_1, ..., a_n, ...)$. Each a_k thus represents some algorithmic or functional composition of 'internal' and 'external' signals.

This path is fed into a highly nonlinear, but otherwise similarly unspecified, decision oscillator which generates an output $h(x)$ that is an element of one of two (presumably) disjoint sets B_0 and B_1 of possible system responses. Let

$$B_0 \equiv b_0, ..., b_k,$$

$$B_1 \equiv b_{k+1}, ..., b_m.$$

Assume a graded response, supposing that if

$$h(x) \in B_0$$

the pattern is not recognized, and if

$$h(x) \in B_1$$

the pattern is recognized and some action $b_j, k + 1 \leq j \leq m$ takes place.

The principal objects of interest are paths x which trigger pattern recognition-and-response exactly once. That is, given a fixed initial state a_0, such that $h(a_0) \in B_0$, we examine all possible subsequent paths x beginning with a_0 and leading exactly once to the event $h(x) \in B_1$. Thus $h(a_0, ..., a_j) \in B_0$ for all $j < m$, but $h(a_0, ..., a_m) \in B_1$.

For each positive integer n, let $N(n)$ be the number of high probability 'grammatical' and 'syntactical' paths of length n which begin with some particular a_0 having $h(a_0) \in B_0$ and lead to the condition $h(x) \in B_1$. Call such paths 'meaningful', assuming, not unreasonably, $N(n)$ to be considerably less than the number of all possible paths of length n leading from a_0 to the condition $h(x) \in B_1$.

While convolution algorithm, the form of the nonlinear oscillator, and the details of grammar and syntax, may all be unspecified in this model, the critical assumption which permits inference on necessary conditions is that the finite limit

$$H \equiv \lim_{n \to \infty} \frac{\log[N(n)]}{n}$$

(3.1)

both exists and is independent of the path x.

We will – not surprisingly – call such a pattern recognition-and-response cognitive process *ergodic*. Not all cognitive processes are likely to be ergodic, implying that H, if it indeed exists at all, is path dependent, although extension to 'nearly' ergodic processes is straightforward.

Invoking the spirit of the Shannon-McMillan Theorem, it is now possible to define an ergodic information source **X** associated with stochastic variates X_j

having joint and conditional probabilities $P(a_0, ..., a_n)$ and $P(a_n|a_0, ..., a_{n-1})$ such that appropriate joint and conditional Shannon uncertainties satisfy the relations

$$H[\mathbf{X}] = \lim_{n \to \infty} \frac{\log[N(n)]}{n} =$$

$$\lim_{n \to \infty} H(X_n|X_0, ..., X_{n-1}) =$$

$$\lim_{n \to \infty} \frac{H(X_0, ..., X_n)}{n}.$$

This information source is defined as *dual* to the underlying ergodic cognitive process.

The Shannon uncertainties $H(...)$, to reiterate, are cross-sectional law-of-large-numbers sums of the form $-\sum_k P_k \log[P_k]$, where the P_k constitute a probability distribution. See Ash (1990) or Cover and Thomas (1991) for the full details, some of which are given in Chapter 2.

The argument constructs a statistical model of simple cognition which, in a somewhat counterintuitive fashion, is similar in spirit to the General Linear Model (GLM) so familiar to researchers. The base, however, is the Shannon-McMillan, rather than the Central Limit, Theorem. As with the GLM, not all phenomena of interest are going to fit.

Treatment of dynamic threshold behavior in consciousness requires iterating the model in much the same sense that a hierarchical linear model represents an iteration of simple or multiple regression (Byrk and Raudenbusch, 2001).

Again, for non-ergodic information sources, a function, $h(x_n)$, of each path $x_n \to x$, may be defined, such that $\lim_{n \to \infty} h(x_n) = h(x)$, but h will not in general be given by the simple cross-sectional laws-of-large numbers analogs above.

Let $s \equiv d(x, \hat{x})$ for high probability paths x and $\hat{x}$, where d is a distortion measure. For 'nearly' ergodic systems one might use something of the form

$$h(\hat{x}) \approx h(x) + sdh/ds|_{s=0}$$

for s sufficiently small. Loosely speaking, the idea is to take a distortion measure as a kind of Finsler metric, imposing a resulting 'global' structure over an appropriate class of non-ergodic information sources. One possible interesting theorem, then, obviously revolves around what properties are metric-independent, in much the same manner as the Rate Distortion Theorem.

This heuristic sketch can be made more precise as follows:

Take a set of 'high probability' paths $x_n \to x$.

Suppose, for all such x, there is an open set, U, containing x, on which the following conditions hold:

(i) For all paths $\hat{x}_n \to \hat{x} \in U$, a distortion measure $s_n \equiv d_U(x_n, \hat{x}_n)$ exists.

(ii) For each path $x_n \to x$ in U there exists a pathwise invariant function $h(x_n) \to h(x)$, in the sense of Khinchin (1957, p.72). While such a function will almost always exist, only in the case of an ergodic information source can it be identified as an 'entropy' in the usual sense.

(iii) A function $F_U(s_n, n) \equiv f_n \to f$ exists, for example, $F_n = s_n, \log[s_n]/n, s_n/n$, and so on.

(iv) The limit

$$\lim_{n \to \infty} \frac{h(x_n) - h(\hat{x}_n)}{f_n} \equiv \nabla_F h|_x$$

(3.2)

exists and is finite.

Under such conditions, various nontrivial standard global atlas/manifold constructions are possible. To reiterate, h is not simply given by the expressions of eq. (2.14) if the source is not ergodic, and the phase transition development of subsequent chapters may be correspondingly more complicated. Restriction to high probability paths simplifies matters considerably, although precisely characterizing them may be difficult, requiring a nontrivial extension of the Shannon-McMillan Theorem.

Different language-analogs will, of course, be defined by different divisions of the total universe of possible responses into different pairs of sets B_0 and B_1, or by requiring more than one response in B_1 along a path. However, like the use of different distortion measures in the Rate Distortion Theorem, it seems obvious that the underlying dynamics will all be qualitatively similar.

Similar but not identical, and herein lies the first of several essential matters: dividing the full set of possible responses into sets B_0 and B_1 may itself require higher order cognitive decisions by another module or modules, suggesting the necessity of 'choice' within a more or less broad set of possible languages-of-thought. This would, in one way, reflect the need of the organism to shift gears according to the different challenges it faces, leading to a model for autocognitive disease when a normally excited state is recurrently (and incorrectly) identified as a member of the 'resting' set B_0, a matter we explore at some length in the last chapter.

A critical complication is that the B-structure is highly *extensible*, since consciousness is well understood to enable an active cognitive learning which expands the possible behavioral repertoire. Think of learning to ride a bicycle: the first few times on the machine require concentrated, conscious attention for every moment and movement, until the skills become unconscious and 'automatic,' at which time the conscious/cognitive decision becomes 'where will I ride today?,' a higher order process. Thus the B-world is itself hierarchical, extensible, and open-ended, an important matter we cannot yet pursue adequately.

Another possible source of structure, however, lies at the input rather than the output end of the model: i.e. the classification of paths instead of outputs. It is possible to define equivalence classes in convolutional 'path space' according to whether a state a_k can be connected by a path with some originating state a_M: in turn, set each possible state to an a_0, and define other states as formally equivalent to it if they can be reached from that (now variable) $a_0 = a_M$ by some real path. That is, a state which can be reached by a legitimate grammatical and syntactical path from a_M is taken as equivalent to it.

Path space can, then, be divided into (ordinarily) disjoint sets of equivalence classes. Each equivalence class defines its own language-of-thought: disjoint cognitive modules, possibly associated with an embedding equivalence class algebra roughly analogous to the standard orbit equivalence construction for dynamical systems. Here, however, are the extraordinarily rich dynamics possible to generalized languages rather than the constrained behavior of the usual distorted, noisy, clock-like contrivances of dynamical systems theory. The image which comes to mind is comparing the often contingent, site-specific genome-environment interaction of evolutionary process with the Newtonian dynamics of a planetary system.

The natural algebraic structure arising from this kind of decomposition is the groupoid (Weinstein, 1996; Brown, 1987).

An open – and important – question is how path algebra structures might relate to B-set structures, particularly given the expanding, hierarchical nature of the latter, suggesting some kind of 'dynamical groupoid' process.

While meaningful paths – creating an inherent grammar and syntax – are defined entirely in terms of system response, as Atlan and Cohen (1998) propose, a critical task is to make these (relatively) disjoint cognitive modules interact, and to examine the effects of that interaction on global properties. One way this can be done is through measures of mutual information and their appropriate asymptotic limit theorems. Invoking the obvious homology with free energy density of a physical system then gives punctuated phase transition in the interaction between modules in what we claim to be a natural manner.

Glazebrook (personal communication) has remarked that this construction can be pieced together up to a global holonomy Lie groupoid using the Aof-Brown globalization theorem (Aof and Brown, 1992), a procedure which might shed further light on the problem of interacting cognitive modules.

The next step is to parametize the information source uncertainty of the dual information source with respect to one or more variates, writing $H[\mathbf{K}]$, where $\mathbf{K} \equiv (K_1, ..., K_s)$ represents a vector in a parameter space. Let the vector $\mathbf{K}$ follow some path in time, tracing out a generalized line or surface $\mathbf{K}(t)$. Following the argument of Wallace (2002b), assume that the probabilities defining H, for the most part, closely track changes in $\mathbf{K}(t)$, so that along a particular 'piece' of a path in parameter space the information source remains as close to memoryless and ergodic as is needed for the mathematics to work. Between pieces, the essential trick will be to impose phase transition characterized by a renormalization symmetry, in the sense of Wilson (1971).

Such an information source will be called 'adiabatically piecewise memoryless ergodic' (APME).

To anticipate the argument, iterating the analysis on paths of 'tuned' sets of renormalization parameters gives a second order punctuation in the rate at which primary interacting information sources representing cognitive submodules become linked to each other: the shifting dynamic workspace of consciousness. The resulting model is, then, to cognition what the HLM is to the GLM in regression theory (Byrk and Raudenbusch, 2001).

Again, significant extension of the model, e.g. to 'mildly' non-ergodic cognition, seems possible, using the atlas/manifold argument implicit to eq. (3.2).

2.   Two neural network examples

Next are two applications of the first order theory to neural networks.

First the simple Hopfield/Hebb stochastic neuron: A series of inputs $y_i^j, i = 1...m$ from m nearby neurons at time j is convoluted with 'weights' $w_i^j, i = 1...m$, using an inner product

$$a_j = \mathbf{y}^j \cdot \mathbf{w}^j = \sum_{i=1}^{m} y_i^j w_i^j$$

(3.3)

in the context of a 'transfer function' $f(\mathbf{y}^j \cdot \mathbf{w}^j)$ such that the probability of the neuron firing and having a discrete output $z^j = 1$ is $P(z^j = 1) = f(\mathbf{y}^j \cdot \mathbf{w}^j)$. Thus the probability that the neuron does not fire at time j is $1 - f(\mathbf{y}^j \cdot \mathbf{w}^j)$.

In the terminology of this chapter the m values y_i^j constitute 'sensory activity' and the m weights w_i^j the 'ongoing activity' at time j, with $a_j = \mathbf{y}^j \cdot \mathbf{w}^j$ and $x = a_0, a_1, ...a_n, ...$

A little more work leads to a fairly standard neural network model in which the network is trained by appropriately varying the **w** through least squares or other error minimization feedback. This can be shown to, essentially, replicate rate distortion arguments, as we can use the error definition to define a distortion function $d(y, \hat{y})$ which measures the difference between the training pattern y and the network output $\hat{y}$ as a function of, for example, the inverse number of training cycles, K. As discussed in some detail elsewhere (Wallace, 2002), learning plateau behavior follows as a phase transition on the parameter K in the mutual information $I(Y, \hat{Y})$.

Park et al. (2000) treat the stochastic neural network in terms of a space of related probability density functions $[p(\mathbf{x}, \mathbf{y}; \mathbf{w}) | \mathbf{w} \in \mathcal{R}^m]$, where **x** is the input, **y** the output and **w** the parameter vector. The goal of learning is to find an optimum $\mathbf{w}^*$ which maximizes the log likelihood function. They define a loss function of learning as

$$L(\mathbf{x}, \mathbf{y}; \mathbf{w}) \equiv -\log p(\mathbf{x}, \mathbf{y}; \mathbf{w}),$$

(3.4)

and one can take as a learning paradigm the gradient relation

$$\mathbf{w}_{t+1} = \mathbf{w}_t - \eta_t \partial L(\mathbf{x}, \mathbf{y}; \mathbf{w}) / \partial \mathbf{w},$$

(3.5)

where η_t is a learning rate.

Park et al. (2000) attack this optimization problem by recognizing that the space of $p(\mathbf{x}, \mathbf{y}; \mathbf{w})$ is Riemannian with a metric given by the Fisher information matrix

$$G(\mathbf{w}) = \int \int \partial \log p / \partial \mathbf{w} [\partial \log p / \partial \mathbf{w}]^T p(\mathbf{x}, \mathbf{y}; \mathbf{w}) \, d\mathbf{y} \, d\mathbf{x},$$

(3.6)

where T is the transpose operation. A Fisher-efficient on-line estimator is then obtained by using the 'natural' gradient algorithm

$$\mathbf{w}_{t+1} = \mathbf{w}_t - \eta_t G^{-1} \partial L(\mathbf{x}, \mathbf{y}; \mathbf{w}) / \partial \mathbf{w}.$$

(3.7)

Again, through the synergistic family of probability distributions $p(\mathbf{x}, \mathbf{y}; \mathbf{w})$, this can be viewed as a special case – a 'representation', to use physics jargon – of the general 'convolution argument' given above.

It seems likely that a rate distortion analysis of the interaction between training language and network response language will nonetheless show the ubiquity of learning plateaus, even in this special case.

Dimitrov and Miller (2001) provide a similar, and very elegant, information-theoretic approach to neural coding and decoding, without, however, addressing punctuation.

3.   Interacting cognitive modules

Two (relatively) distinct cognitive submodules can be represented by two distinct sequences of states, the convolutional paths $x \equiv x_0, x_1, \ldots$ and $y \equiv y_0, y_1, \ldots$. These paths are, however, both very highly structured and serially correlated and have dual information sources $\mathbf{X}$ and $\mathbf{Y}$. Since the modules, in reality, interact through some kind of endless back-and-forth mutual crosstalk, these sequences of states are not independent, but are jointly serially correlated. We can, then, define a path of sequential pairs as $z \equiv (x_0, y_0), (x_1, y_1), \ldots$. The essential content of the Joint Asymptotic Equipartition Theorem (JAEPT), a variant of the Shannon-McMillan Theorem, is that the set of joint paths z can be partitioned into a relatively small set of high probability termed *jointly typical*, and a much larger set of vanishingly small probability. Further, according to the JAEPT, the *splitting criterion* between high and low probability sets of pairs is the mutual information

$$I(X, Y) = H(X) - H(X|Y) = H(X) + H(Y) - H(X, Y),$$

where $H(X), H(Y), H(X|Y)$ and $H(X, Y)$ are, respectively, the (cross-sectional) Shannon uncertainties of X and Y, their conditional uncertainty, and their joint uncertainty. See Cover and Thomas (1991) for mathematical details.

Again, similar approaches to neural process have been recently adopted by Dimitrov and Miller (2001).

Note that, using this asymptotic limit theorem approach, one need not model the exact form or dynamics of the crosstalk feedback. Crushing algebraic complexities can be postponed until a later stage of the argument. They will, however, appear in due course with some vengeance.

The high probability pairs of paths are, in this formulation, all equiprobable, and if $N(n)$ is the number of jointly typical pairs of length n, then

$$I(X, Y) = \lim_{n \to \infty} \frac{\log[N(n)]}{n}.$$

Extending the earlier language-on-a-network models of Wallace and Wallace (1998, 1999), we suppose there is a coupling parameter P representing the degree of linkage between the modules, and set $K = 1/P$, following the development of those earlier studies. Note that in a brain model this parameter represents the intensity of coupling between distant neural structures.

Then

$$I[K] = \lim_{n \to \infty} \frac{\log[N(K, n)]}{n}.$$

The essential 'homology' between information theory and statistical mechanics lies in the similarity of this expression with the infinite volume limit of the free energy density. If $Z(K)$ is the statistical mechanics partition function derived from the system's Hamiltonian, then the free energy density is determined by the relation

$$F[K] = \lim_{V \to \infty} \frac{\log[Z(K)]}{V}.$$

(3.8)

F is the free energy density, V the system volume and $K = 1/T$, where T is the system temperature.

As described at the end of Chapter 2, imposition of this homology permits importation of renormalization methods into information theory. Imposition of invariance under renormalization on the mutual information splitting criterion $I(X, Y)$ implies the existence of phase transitions analogous to learning plateaus or punctuated evolutionary equilibria. The next chapter gives an extensive development.

The physiological details of mechanism, in this model, are captured by the definitions of coupling parameter, renormalization symmetry, and, perhaps, the distribution of the renormalization across agency, a matter we treat below.

Here, however, these changes are perhaps better described as 'punctuated interpenetration' between interacting cognitive modules.

Detailed dynamics depend on the choice of renormalization symmetry and distribution, which are likely to reflect particularities of mechanism – the manner in which the dynamics of the forest are dependent on the physiology of individual trees, albeit in a many-to-one manner. Renormalization in cognitive structures is not likely to follow simple physical analogs, and may well be subject, in addition to complications of distribution, to the 'tuning' of universality class parameters that are characteristically fixed for simple physical systems. The algebra is straightforward if complicated, and given in the following chapter.

Chapter 4

THE FLUCTUATING DYNAMIC THRESHOLD

1. Language-on-a-network models

Earlier work (Wallace and Wallace, 1998; 1999) addressed how a language, in a large sense, 'spoken' on a network structure, responds as properties of the network change. The language might be speech, pattern recognition, or cognition. The network might be social, chemical, or neural. The properties of interest were the magnitude of 'strong' or 'weak' ties which, respectively, either disjointly partitioned the network or linked it across such partitioning. These would be analogous to local and mean-field couplings in physical systems.

Fix the magnitude of strong ties – again, those which disjointly partition the underlying network into cognitive or other submodules – but vary the index of nondisjunctive weak ties, P, between components, taking $K = 1/P$.

Assume the piecewise, adiabatically memoryless ergodic information source (or sources) dual to cognitive process depends on three parameters, two explicit and one implicit. The explicit are K as above and, as a calculational device, an 'external field strength' analog J, which gives a 'direction' to the system. We will, in the limit, set $J = 0$. Note that many other approaches may well be possible, since renormalization techniques are more philosophy than prescription.

The implicit parameter, r, is an inherent generalized 'length' characteristic of the phenomenon, on which J and K are defined. That is, J and K are written as functions of averages of the parameter r, which may be quite complex, having nothing at all to do with conventional ideas of space. For example r may be defined by the degree of niche partitioning in ecosystems or separation in social structures.

For a given generalized language of interest having a well defined (adiabatically, piecewise memoryless) ergodic source uncertainty, $H = H[K, J, \mathbf{X}]$.

To summarize a long train of standard argument (Binney et al., 1986; Wilson, 1971), imposition of invariance of H under a renormalization transform in the implicit parameter r leads to expectation of both a critical point in K, written K_C, reflecting a phase transition to or from collective behavior across the entire array, and of power laws for system behavior near K_C. Addition of other parameters to the system results in a 'critical line' or surface.

Let $\kappa \equiv (K_C - K)/K_C$ and take χ as the 'correlation length' defining the average domain in r-space for which the information source is primarily dominated by 'strong' ties. The first step is to average across r-space in terms of 'clumps' of length $R = <r>$. Then $H[J, K, \mathbf{X}] \rightarrow H[J_R, K_R, \mathbf{X}]$.

Taking Wilson's (1971) analysis as a starting point – not the only way to proceed – the 'renormalization relations' used here are:

$$H[K_R, J_R, \mathbf{X}] = f(R)H[K, J, \mathbf{X}]$$

$$\chi(K_R, J_R) = \frac{\chi(K, J)}{R},$$

(4.1)

with $f(1) = 1$ and $J_1 = J, K_1 = K$. The first equation significantly extends Wilson's treatment. It states that 'processing capacity,' as indexed by the source uncertainty of the system, representing the 'richness' of the generalized language, grows monotonically as $f(R)$, which must itself be a dimensionless function in R, since both $H[K_R, J_R]$ and $H[K, J]$ are themselves dimensionless. Most simply, this requires replacing R by R/R_0, where R_0 is the 'characteristic length' for the system over which renormalization procedures are reasonable, then setting $R_0 \equiv 1$, hence measuring length in units of R_0.

Wilson's original analysis focused on free energy density. Under 'clumping,' densities must remain the same, so that if $F[K_R, J_R]$ is the free energy of the clumped system, and $F[K, J]$ is the free energy density before clumping, then Wilson's equation (4) is $F[K, J] = R^{-3}F[K_R, J_R]$,

$$F[K_R, J_R] = R^3 F[K, J].$$

Remarkably, the renormalization equations are solvable for a broad class of functions $f(R)$, or more precisely, $f(R/R_0), R_0 \equiv 1$.

The second equation just states that the correlation length simply scales as R.

Again, the central feature of renormalization in this context is the assumption that, at criticality, the system looks the same at all scales, that is, it is *invariant under renormalization* at the critical point. All else flows from this.

There is no unique renormalization procedure for information sources: other, very subtle, symmetry relations – not necessarily based on the elementary physical analog we use here – may well be possible. For example, McCauley (1993, p.168) describes the highly counterintuitive renormalizations needed to understand phase transition in simple 'chaotic' systems. This is important, since biological or social systems may well alter their renormalization properties – equivalent to tuning their phase transition dynamics – in response to external signals. We will make much use of a simple version of this possibility, termed 'universality class tuning,' below.

To begin, following Wilson, take $f(R) = R^d$, d some real number $d > 0$, and restrict K to near the 'critical value' K_C. If $J \rightarrow 0$, a simple series expansion and some clever algebra (Wilson, 1971; Binney et al., 1986) gives

$$H = H_0 \kappa^\alpha$$

$$\chi = \frac{\chi_0}{\kappa^s},$$

(4.2)

where α, s are positive constants. More biologically relevant examples appear below.

As Dimitrov pointed out in reviewing an early version of this book, critical behavior in information systems has been studied elsewhere. Both the information bottleneck work of Tishby et al. (1999) and the generalization by Parker et al. (2003, fig. 1) observe bifurcations which Tishby et al. (1999) in fact describe as "[bifurcating] at some finite (critical) [parameter], through a second-order phase transition... [forming] a hierarchy of relevant quantizations..." Both groups treat such phenomena as sidelights to an optimization calculation, while here we take them as central to the enterprise.

Further from the critical point, matters are more complicated, appearing to involve Generalized Onsager Relations and a kind of thermodynamics associated with a Legendre transform of H: $S \equiv H - K dH/dK$ (Wallace, 2002a). Although this extension is quite important to describing behaviors away from criticality, the mathematical detail is cumbersome. A more detailed discussion appears at the end of this chapter.

An essential insight is that *regardless of the particular renormalization prop-erties, sudden critical point transition is possible in the opposite direction for this model.* That is, going from a number of independent, isolated and frag-mented systems operating individually and more or less at random, into a single large, interlocked, coherent structure, once the parameter K, the inverse strength of weak ties, falls below threshold, or, conversely, once the strength of weak ties parameter $P = 1/K$ becomes large enough.

Thus, increasing nondisjunctive weak ties between them can bind several different cognitive 'language' functions into a single, embedding hierarchical metalanguage containing each as a linked subdialect, and do so in an inherently punctuated manner. This could be a dynamic process, creating a shifting, ever-changing pattern of linked cognitive submodules, according to the challenges or opportunities faced by the organism.

This heuristic insight can be made more exact using a rate distortion argu-ment (or, more generally, using the Joint Asymptotic Equipartition Theorem) as follows (Wallace, 2002a, b):

Suppose that two ergodic information sources Y and B begin to interact, to 'talk' to each other, to influence each other in some way so that it is possible, for example, to look at the output of B – strings b – and infer something about the behavior of Y from it – strings y. We suppose it possible to define a retranslation from the B-language into the Y-language through a deterministic code book, and call $\hat{Y}$ the translated information source, as mirrored by B.

Define some distortion measure comparing paths y to paths $\hat{y}$, $d(y, \hat{y})$. In-voke the Rate Distortion Theorem's mutual information $I(Y, \hat{Y})$, which is the splitting criterion between high and low probability pairs of paths. Impose, now, a parametization by an inverse coupling strength K, and a renormaliza-tion representing the global structure of the system coupling. This may be much different from the renormalization behavior of the individual components. If $K < K_C$, where K_C is a critical point (or surface), the two information sources will be closely coupled enough to be characterized as condensed.

In the absence of a distortion measure, the Joint Asymptotic Equipartition Theorem gives a similar result.

Detailed coupling mechanisms will be sharply constrained through regular-ities of grammar and syntax imposed by limit theorems associated with phase transition.

Wallace and Wallace (1998, 1999) and Wallace (2002) use this approach to address certain evolutionary processes in a relatively unified fashion. These papers, and those of Wallace and Fullilove (1999) and Wallace (2002a), further describe how biological or social systems might respond to gradients in infor-mation source uncertainty and related quantities when the system is away from phase transition. Language-on-network systems, as opposed to physical sys-tems, appear to diffuse away from concentrations of an 'instability' construct related to a Legendre transform of information source uncertainty, in much the same way entropy is the Legendre transform of free energy density in a physical system.

Simple thermodynamics addresses physical systems held at or near equilibrium. Treatment of nonequilibrium, for example highly dynamic, systems requires significant extension of thermodynamic theory. The most direct approach has been the first-order phenomenological theory of Onsager, which involves relating first order rate changes in system parameters K_j to gradients in physical entropy S, involving 'Onsager relation' equations of the form

$$\sum_k R_{k,j} dK_j/dt = \partial S/\partial K_j,$$

where the $R_{k,j}$ are characteristic constants of a particular system and S is defined to be the Legendre transform of free energy density F;

$$S \propto F - \sum_j K_j \partial F/\partial K_j.$$

The entropy-analog for an information system is, then, the dimensionless quantity

$$S \equiv H - \sum_j K_j \partial H/\partial K_j,$$

or a similar equation in the mutual information I.

Note that in this treatment I or H play the role of free energy, not entropy, and that their Legendre transform plays the role of physical entropy. This is a key matter.

For information systems, a parametized 'instability', $Q[K] \equiv S - H$, is defined from the principal splitting criterion by the relations

$$Q[K] = -KdH[K]/dK$$

$$Q[K] = -KdI[K]/dK,$$

(4.3)

where $H[K]$ and $I[K]$ are, respectively, information source uncertainty or mutual information in the Asymptotic Equipartition, Rate Distortion, or Joint Asymptotic Equipartition Theorems.

Extension of thermodynamic theory to information systems involves a first order system of phenomenological equations analogous to the Onsager relations, but possibly having very complicated behavior in the $R_{j,k}$. These will not necessarily produce simple diffusion toward peaks in S, as would be expected for a simple physical system. For example, as discussed, there is evidence that social network structures are affected by diffusion *away* from concentrations in the S-analog. Thus the phenomenological relations affecting the dynamics of information networks, which are inherently open systems, may not be governed simply by mechanistic diffusion toward 'peaks in entropy', but may, in first order, display far more complicated behavior. We will return to this point repeatedly.

2.   'Biological' phase transitions

Now the mathematical detail concealed by the invocation of the asymptotic limit theorems emerges with a vengeance. Equation (4.1) states that the information source and the correlation length, the degree of coherence on the underlying network, scale under renormalization clustering in chunks of size R as

$$H[K_R, J_R]/f(R) = H[J, K]$$

$$\chi[K_R, J_R]R = \chi(K, J),$$

with $f(1) = 1, K_1 = K, J_1 = J$, where we have slightly rearranged terms.

Differentiating these two equations with respect to R, so that the right hand sides are zero, and solving for dK_R/dR and dJ_R/dR gives, after some consolidation, expressions of the form

$$dK_R/dR = u_1 d\log(f)/dR + u_2/R$$

$$dJ_R/dR = v_1 J_R d\log(f)/dR + \frac{v_2}{R}J_R.$$

(4.4)

The $u_i, v_i, i = 1, 2$ are functions of K_R, J_R, but not explicitly of R itself.

We expand these equations about the critical value $K_R = K_C$ and about $J_R = 0$, obtaining

$$dK_R/dR = (K_R - K_C)yd\log(f)/dR + (K_R - K_C)z/R$$

$$dJ_R/dR = wJ_Rd\log(f)/dR + xJ_R/R.$$

(4.5)

The terms $y = du_1/dK_R|_{K_R=K_C}, z = du_2/dK_R|_{K_R=K_C}, w = v_1(K_C,0), x = v_2(K_C,0)$ are constants.

Solving the first of these equations gives

$$K_R = K_C + (K - K_C)R^z f(R)^y,$$

(4.6)

again remembering that $K_1 = K, J_1 = J, f(1) = 1$.

Wilson's essential trick is to iterate on this relation, which is supposed to converge rapidly near the critical point (Binney et al., 1986), assuming that for K_R near K_C, we have

$$K_C/2 \approx K_C + (K - K_C)R^z f(R)^y.$$

(4.7)

We iterate in two steps, first solving this for $f(R)$ in terms of known values, and then solving for R, finding a value R_C that we then substitute into the first of equations (4.1) to obtain an expression for $H[K, 0]$ in terms of known functions and parameter values.

The first step gives the general result

$$f(R_C) \approx \frac{[K_C/(K_C - K)]^{1/y}}{2^{1/y} R_C^{z/y}}.$$

(4.8)

Solving this for R_C and substituting into the first expression of equation (4.1) gives, as a first iteration of a far more general procedure (Shirkov and Kovalev, 2001), the result

$$H[K, 0] \approx \frac{H[K_C/2, 0]}{f(R_C)} = \frac{H_0}{f(R_C)}$$

$$\chi(K, 0) \approx \chi(K_C/2, 0)R_C = \chi_0 R_C,$$

(4.9)

which are the essential relationships.

Note that a power law of the form $f(R) = R^m, m = 3$, which is the direct physical analog, may not be biologically reasonable, since it says that 'language richness' can grow very rapidly as a function of increased network size. Such rapid growth is simply not observed.

Taking the biologically realistic example of non-integral 'fractal' exponential growth,

$$f(R) = R^\delta,$$

(4.10)

where $\delta > 0$ is a real number which may be quite small, equation (4.8) can be solved for R_C, obtaining

$$R_C = \frac{[K_C/(K_C - K)]^{[1/(\delta y + z)]}}{2^{1/(\delta y + z)}}$$

(4.11)

for K near K_C. Note that, for a given value of y, one might characterize the relation $\alpha \equiv \delta y + z = $ constant as a 'tunable universality class relation' in the sense of Albert and Barabasi (2002).

Substituting this value for R_C back into equation (4.8) gives a somewhat more complex expression for H than equation (4.1), having three parameters: δ, y, z.

A more biologically interesting choice for $f(R)$ is a logarithmic curve that 'tops out', for example

$$f(R) = m \log(R) + 1.$$

(4.12)

Again $f(1) = 1$.
Using Mathematica 4.2 or above to solve equation (4.8) for R_C gives

$$R_C = [\frac{Q}{LambertW[Q \exp(z/my)]}]^{y/z},$$

(4.13)

where

$$Q \equiv (z/my)2^{-1/y}[K_C/(K_C - K)]^{1/y}.$$

The transcendental function LambertW(x) is defined by the relation

$$LambertW(x) \exp(LambertW(x)) = x.$$

It arises in the theory of random networks and in renormalization strategies for quantum field theories.

An asymptotic relation for $f(R)$ would be of particular biological interest, implying that 'language richness' increases to a limiting value with population growth. Such a pattern is broadly consistent with calculations of the degree of allelic heterozygosity as a function of population size under a balance between genetic drift and neutral mutation (Hartl and Clark, 1997; Ridley, 1996). Taking

$$f(R) = \exp[m(R-1)/R]$$

(4.14)

gives a system which begins at 1 when $R = 1$, and approaches the asymptotic limit $\exp(m)$ as $R \to \infty$. Mathematica 4.2 finds

$$R_C = \frac{my/z}{LambertW[A]},$$

(4.15)

where

$$A \equiv (my/z)\exp(my/z)[2^{1/y}[K_C/(K_C - K)]^{-1/y}]^{y/z}.$$

These developments indicate the possibility of taking the theory significantly beyond arguments by abduction from simple physical models, although the notorious difficulty of implementing information theory existence arguments will undoubtedly persist.

3.   Universality class distribution

Physical systems undergoing phase transition usually have relatively pure renormalization properties, with quite different systems clumped into the same 'universality class,' having fixed exponents at transition (Binney et al., 1986). Biological and social phenomena may be far more complicated:

If the system of interest is a mix of subgroups with different values of some significant renormalization parameter m in the expression for $f(R, m)$, according to a distribution $\rho(m)$, then the first expression in equation (4.1) should generalize, at least to first order, as

$$H[K_R, J_R] = < f(R, m) > H[K, J]$$

$$\equiv H[K, J] \int f(R, m)\rho(m)dm.$$

(4.16)

If $f(R) = 1 + m \log(R)$ then, given any distribution for m,

$$< f(R) > = 1 + < m > \log(R)$$

(4.17)

where $< m >$ is simply the mean of m over that distribution.

Other forms of $f(R)$ having more complicated dependencies on the distributed parameter or parameters, like the power law R^δ, do not produce such a simple result. Taking $\rho(\delta)$ as a normal distribution, for example, gives

$$< R^\delta > = R^{<\delta>} \exp[(1/2)(\log(R^\sigma))^2],$$

(4.18)

where σ^2 is the distribution variance. The renormalization properties of this function can be determined from equation (4.8), and the calculation is left to the reader as an exercise, best done in Mathematica 4.2 or above.

Thus the information dynamic phase transition properties of mixed systems will not in general be simply related to those of a single subcomponent, a matter of possible empirical importance: If sets of relevant parameters defining renormalization universality classes are indeed distributed, experiments observing pure phase changes may be very difficult. Tuning among different possible renormalization strategies in response to external signals would result in even greater ambiguity in recognizing and classifying information dynamic phase transitions.

Important aspects of mechanism may be reflected in the combination of renormalization properties and the details of their distribution across subsystems.

In sum, real biological, social, or interacting biopsychosocial systems are likely to have very rich patterns of phase transition which may not display the simplistic, indeed, literally elemental, purity familiar to physicists. Overall mechanisms will, however, still remain significantly constrained by the theory, in the general sense of probability limit theorems.

4.   Punctuated universality class tuning

The next step is to iterate the general argument onto the process of phase transition itself, producing a model of consciousness as a tunable neural workspace subject to inherent punctuated detection of external events.

As described above, an essential character of physical systems subject to phase transition is that they belong to particular 'universality classes'. Again, this means that the exponents of power laws describing behavior at phase transition will be the same for large groups of markedly different systems, with 'natural' aggregations representing fundamental class properties (Binney et al., 1986).

It appears that biological or social systems undergoing phase transition analogs need not be constrained to such classes, and that 'universality class tuning', meaning the strategic alteration of parameters characterizing the renormalization properties of punctuation, might well be possible. Here we focus on the tuning of parameters within a single, given, renormalization relation. Clearly, however, wholesale shifts of renormalization properties must ultimately be considered as well, a matter for future work.

Universality class tuning has been observed in models of 'real world' networks. As Albert and Barabasi (2002) put it,

> The inseparability of the topology and dynamics of evolving networks is shown by the fact that [the exponents defining universality class] are related by [a] scaling relation..., underlying the fact that a network's assembly uniquely determines its topology. However, in no case are these exponents unique. They can be tuned continuously...

Suppose that a structured external environment, itself an appropriately regular information source Y, 'engages' a modifiable cognitive system. The environment begins to write an image of itself on the cognitive system in a distorted manner permitting definition of a mutual information $I[K]$ splitting criterion according to the Rate Distortion or Joint Asymptotic Equipartition Theorems. K is an inverse coupling parameter between system and environment (Wallace, 2002a, b). At punctuation – near some critical point K_C – the systems begin to interact very strongly indeed, and, near K_C, using the simple physical model of equation (4.2),

$$I[K] \approx I_0 [\frac{K_C - K}{K_C}]^\alpha.$$

For a physical system α is fixed, determined by the underlying 'universality class.' Here we will allow α to vary, and, in the section below, to itself respond explicitly to signals.

Normalizing K_C and I_0 to 1,

$$I[K] \approx (1 - K)^\alpha.$$

(4.19)

The horizontal line $I[K] = 1$ corresponds to $\alpha = 0$, while $\alpha = 1$ gives a declining straight line with unit slope which passes through 0 at $K = 1$. Consideration shows there are progressively sharper transitions between the necessary zero value at $K = 1$ and the values defined by this relation for $0 < K, \alpha < 1$. The rapidly rising slope of transition with declining α is of considerable significance:

The instability associated with the splitting criterion $I[K]$ is defined by

$$Q[K] \equiv -K dI[K]/dK = \alpha K (1 - K)^{\alpha - 1},$$

(4.20)

and is singular at $K = K_C = 1$ for $0 < \alpha < 1$. Following earlier work (Wallace and Wallace, 1998, 1999; Wallace and Fullilove, 1999; Wallace, 2002a), we interpret this to mean that values of $0 < \alpha \ll 1$ are highly unlikely for real systems, since $Q[K]$, in this model, represents a kind of barrier for 'social' information systems, in particular interacting neural network modules, a matter explored further below.

On the other hand, smaller values of α mean that the system is far more efficient at responding to the adaptive demands imposed by the embedding structured environment, since the mutual information which tracks the matching of internal response to external demands, $I[K]$, rises more and more quickly toward the maximum for smaller and smaller α as the inverse coupling parameter K declines below $K_C = 1$. That is, systems able to attain smaller α are more responsive to external signals than those characterized by larger values, in this model, but smaller values will be harder to reach, probably only at some considerable physiological or opportunity cost. Focused conscious action takes resources, of one form or another.

A subsequent chapter makes these considerations explicit, modeling the role of contextual and energy constraints on the relations between Q, I, and other system properties.

The more biologically realistic renormalization strategies given above produce sets of several parameters defining the universality class, whose tuning gives behavior much like that of α in this simple example.

Formal iteration of the phase transition argument on this calculation gives tunable consciousness, focusing on paths of universality class parameters.

Suppose the renormalization properties of a language-on-a network system at some 'time' k are characterized by a set of parameters $A_k \equiv \alpha_1^k, ..., \alpha_m^k$. Fixed parameter values define a particular universality class for the renormalization. We suppose that, over a sequence of 'times,' the universality class properties can be characterized by a path $x_n = A_0, A_1, ..., A_{n-1}$ having significant serial correlations which, in fact, permit definition of an adiabatically piecewise memoryless ergodic information source associated with the paths x_n. We call that source $\mathbf{X}$.

Suppose also, in the now-usual manner, that the set of external (or internal, systemic) signals impinging on consciousness is also highly structured and forms another information source $\mathbf{Y}$ which interacts not only with the system of interest globally, but specifically with its universality class properties as characterized by $\mathbf{X}$. $\mathbf{Y}$ is necessarily associated with a set of paths y_n.

Pair the two sets of paths into a joint path, $z_n \equiv (x_n, y_y)$ and invoke an inverse coupling parameter, K, between the information sources and their paths. This leads, by the arguments above, to phase transition punctuation of $I[K]$, the mutual information between $\mathbf{X}$ and $\mathbf{Y}$, under either the Joint Asymptotic Equipartition Theorem or under limitation by a distortion measure, through the Rate Distortion Theorem. The essential point is that $I[K]$ is a splitting criterion under these theorems, and thus partakes of the homology with free energy density which we have invoked above.

Activation of universality class tuning, the model's version of attentional focusing, then becomes itself a punctuated event in response to increasing linkage between the organism and an external structured signal or some particular system of internal events.

This iterated argument exactly parallels the extension of the General Linear Model to the Hierarchical Linear Model in regression theory (Byrk and Raudenbusch, 2001).

Another path to the fluctuating dynamic threshold might be through a second order iteration similar to that just above, but focused on the parameters defining the universality class distributions of section 4.3.

Following recent arguments of Gillooly et al. (2004) showing metabolic rate can calibrate the molecular clock of evolutionary process, and taking into account the crude analogy between punctuated equilibrium in evolutionary, and learning plateaus in cognitive, systems (Wallace, 2002b), it seems likely that the generalized Onsager relation arguments used above can be iterated as well. As we will show in Chapter 5, such iteration must take place in the context of competing energy constraints defined at different levels of organization.

The question becomes, as in the study of the idiotypic networks of immune function, the rate of convergence of the iterative process. It seems likely that a small number of iterations will suffice to explain most current controlled experiments.

More generally, the development of Wallace (2003) suggests the possibility of one or more tunable internal 'retinas' for cognitive process which could be adjusted to accelerate convergence. This idea has certain interesting implications, which are explored in the next chapter.

5.   Dynamics far from criticality

Attention has thus far primarily focused on the dynamic properties of a parametized information source near a critical surface. We now ask in more detail how a parametized information source behaves 'normally', far from such a surface.

To reiterate, according to the Shannon-McMillan Theorem, the number $N(n)$ of meaningful paths of length n emitted by an information source satisfies $H[\mathbf{X}] = \lim_{n \to \infty} \log[N(n)]/n$, where $H[\mathbf{X}]$ is the source uncertainty defined from the joint and conditional probabilities of the paths x.

For a physical system the free energy density is defined by the analogous relation:

$$F(K_1, ..., K_m) = \lim_{V \to \infty} \frac{\log[Z(K_1, ..., K_m, V)]}{V}$$

where V is the system volume, $K_1, ..., K_m$ are other system-wide parameters, and $Z(K_1, ..., K_m, V)$ is the partition function defined from system energy states, and K_1 is an inverse temperature.

Previously, this homology was used to impose renormalization symmetries relating 'phase change' to underlying architecture for parametized information sources. Here, further use of it, far from critical points or surfaces, connects architecture and dynamics, in a large sense.

Parametize the information source uncertainty of interest, so that $H = H[K_1, ..., K_m, \mathbf{X}] = H[\mathbf{K}, \mathbf{X}]$ where now K_1 as the inverse of the 'strength of weak ties' across some structure.

For a physical system the equation of state which describes the macroscopic behavior of the system emerges through imposition of a *Legendre transform* on free energy. The Legendre transform of a well-behaved function $f(K_1, ..., K_m)$ is defined by

$$g = f - \sum_{i=1}^{w} K_i \partial f / \partial K_i$$

$$\equiv f - \sum_{i=1}^{w} K_i Q_i,$$

(4.21)

so that $Q_i = \partial f / \partial K_i$, and is invertible provided $\partial f / \partial K$ is well behaved. Then

$$f = g - \sum_{i=1}^{w} Q_i \partial g / \partial Q_i.$$

(4.22)

The generalization when f is not well-behaved is through a variational argument (Fredlin and Wentzell, 1998; Dembo and Zeitouni, 1998) rather than this tangent plane argument.

In a physical system for which F is the free energy density, the Legendre transform defines the macroscopic entropy as

$$S \equiv F - \sum_{i} K_i \partial F / \partial K_i.$$

(4.23)

The associated generalized forces constituting the equation(s) of state are then

$$Q_i \equiv \partial F/\partial K_i.$$

The analogous argument generates a series of macroscopic equations of state for a system characterized by a parametized information source uncertainty $H[\mathbf{K}; \mathbf{X}]$:

$$S \equiv H - \sum_i K_i \partial H/\partial K_i = H - \mathbf{K} \cdot \nabla|_{\mathbf{K}} H$$

$$Q_i \equiv \partial H/\partial K_i,$$

(4.24)

where S is now defined as the macroscopic *disorder* and $\Omega \equiv S - H$ the *instability*.

If the rates of change of the K_i have independent effect, then we would write

$$S \equiv H - (\mathbf{K}, \dot{\mathbf{K}}) \cdot \nabla H|_{(\mathbf{K}, \dot{\mathbf{K}})},$$

and similarly generalize the definition of the Q_i.

Note in particular that each hierarchical ordering relation will add a new 'strength of weak ties' parameter to the thermodynamics.

For physical systems the 'Onsager relations' define the system's response to entropy. These assume that time rates of change of the defining parameters, the K_i, are in direct proportion to 'thermodynamic forces' defined as gradients in the entropy with respect to the characteristic parameters:

$$dK_i/dt = \sum_j L_{i,j} \partial S/\partial K_j.$$

Next we treat the opposite case in more detail; that is, similar to the usual physics treatment, with one significant change. Assume there is an internal 'social' structure across the coupled assembly that indeed responds to gradients in the disorder construct S, but with opposite sign to that of a physical system: A 'social' system is seen to move *away* from concentrations of 'disorder' rather than towards it. Thus we have, for the m parameters of $H[K_1, K_2, ...K_m, \mathbf{X}]$, m equations of a 'generalized Onsager relation'

$$dK_i/dt = -L\partial S/\partial K_i,$$

(4.25)

where L is positive. Next, adjust the system to some initial reference configuration $K_0 = K_1, K_2, ...K_m$ such that $dK/dt|_{K_0} = -L\nabla S|_{K_0} \equiv 0$.

Deviations from this reference configuration, $\delta K \equiv K - K_0$, in first order, obey the relation

$$d\delta K_i/dt \approx -L\sum_{j=1}^{m}(\partial^2 S/\partial K_i\partial K_j|_{K_0})\delta K_j.$$

(4.26)

In matrix form, writing $U_{i,j} = \partial^2 S/\partial K_i\partial K_j = U_{j,i}$, this becomes

$$d\delta K/dt = -LU\delta K.$$

(4.27)

Assume the appropriate regularity conditions on S and $\mathbf{U}$ and expand the deviations vector δK in terms of the eigenvectors of the symmetric matrix $\mathbf{U}$, m-dimensional vectors e_i such that $\mathbf{U}e_i = \lambda_i e_i$, so that

$$\delta K = \sum_{i=1}^{m}\delta a_i e_i.$$

(4.28)

Equation (4.27) then has the solution

$$\delta K(t) = \sum_{i=1}^{m} \delta a_i \exp(-L\lambda_i t) e_i.$$

(4.29)

Clearly any eigenconfiguration e_j having a negative eigenvalue, $\lambda_j < 0$, will be amplified exponentially in time until the network goes into precisely the sudden epileptiform phase transition studied in Wallace (2000). Clearly, for a given reference configuration K_0, different e_j with different negative eigenvalues will lead to slightly different forms of epileptiform detection transitions.

This approach, organized around a 'tunable reference configuration K_0, will be formalized in the next chapter as a generalized (tunable) retina.

Continuing the theoretical development another step, let $d\delta K_i/dt \equiv \delta V_i$, and rewrite the first order approximation to obtain an expression for the magnitude of $(\delta V)^2$:

$$(\delta V)^2 \approx L^2/2 \sum_{i,j} [\sum_k U_{i,k} U_{k,j}] \delta K_i \delta K_j.$$

(4.30)

Defining

$$g_{i,j} \equiv L^2/2 \sum_k (\partial^2 S/\partial K_i \partial K_k)(\partial^2 S/\partial K_k \partial K_j)$$

(4.31)

produces something much like the fundamental relation of a Riemannian differential geometry:

$$dV^2 = \sum_{i,j} g_{i,j}(K)dK_i dK_j.$$

(4.32)

A geodesic represents, in this configuration, not a minimization of V along some path in K-space, but rather its maximization, equivalent to a minimization of the time-of-flight. The model is of a ball bearing rolling down a hill. A quasi-stochastic extension would see a Brownian fuzz around optimal paths. Not unexpectedly, the argument has recovered something much like the results of the purely physical treatment, albeit with reversed sign.

Chapter 5

EXTENDING THE MODEL

1. The simplest tunable retina

The iterated development of the Section 4.4 – analogous to expanding the GLM to the HLM – which involved paths in renormalization-parameter space, can itself be significantly extended. This produces a generalized tunable retina model which can be interpreted as a 'Rate Distortion manifold,' a concept which further opens the way for import of a vast array of tools from geometry and topology.

Suppose, now, that threshold behavior in conscious reaction requires some elaborate system of nonlinear relationships defining the set of renormalization parameters $A_k \equiv \alpha_1^k, ..., \alpha_m^k$ above. The critical assumption is that there is a tunable 'zero order state,' and that changes about that state are, in first order, relatively small, although their effects on punctuated process may not be at all small. Thus, given an initial m-dimensional vector A_k, the parameter vector at time $k + 1$, A_{k+1}, can, in first order, be written as

$$A_{k+1} \approx \mathbf{R}_{k+1} A_k,$$

(5.1)

where $\mathbf{R}_{t+1}$ is an $m \times m$ matrix, having m^2 components.

If the initial parameter vector at time $k = 0$ is A_0, then at time k

$$A_k = \mathbf{R}_k \mathbf{R}_{k-1} ... \mathbf{R}_1 A_0.$$

(5.2)

The interesting correlates of consciousness are, in this development, *now represented by an information-theoretic path defined by the sequence of operators* $\mathbf{R}_k$, each member having m^2 components. The grammar and syntax of the path defined by these operators is associated with a dual information source, in the usual manner.

The effect of an information source of external signals, $\mathbf{Y}$ in the discussion above, is now seen in terms of more complex joint paths in Y and R-space whose behavior is, again, governed by a mutual information splitting criterion according to the JAEPT.

The complex sequence in m^2-dimensional R-space has, by this construction, been projected down onto a parallel path, the smaller set of m-dimensional α-parameter vectors $A_0, ..., A_k$.

If the punctuated tuning of consciousness is now characterized by a 'higher' dual information source – an embedding generalized language – so that the paths of the operators $\mathbf{R}_k$ are autocorrelated, then the autocorrelated paths in A_k represent output of a parallel information source which is, given Rate Distortion limitations, apparently a grossly simplified, and hence highly distorted, picture of the 'higher' conscious process represented by the R-operators, having m as opposed to $m \times m$ components.

High levels of distortion may not necessarily be the case for such a structure.

Let us examine a single iteration in more detail, assuming now there is a (tunable) zero reference state, $\mathbf{R}_0$, for the sequence of operators $\mathbf{R}_k$, and that

$$A_{k+1} = (\mathbf{R}_0 + \delta \mathbf{R}_{k+1})A_k,$$

(5.3)

where $\delta \mathbf{R}_k$ is 'small' in some sense compared to $\mathbf{R}_0$.

Note that in this analysis the operators $\mathbf{R}_k$ are, implicitly, determined by linear regression. We thus can invoke a quasi-diagonalization in terms of $\mathbf{R}_0$. Let $\mathbf{Q}$ be the matrix of eigenvectors which Jordan-block-diagonalizes $\mathbf{R}_0$. Then

$$QA_{k+1} = (QR_0Q^{-1} + Q\delta R_{k+1}Q^{-1})QA_k.$$

(5.4)

If QA_k is an eigenvector of R_0, say Y_j with eigenvalue λ_j, it is possible to rewrite this equation as a generalized spectral expansion

$$Y_{k+1} = (J + \delta J_{k+1})Y_j \equiv \lambda_j Y_j + \delta Y_{k+1}$$

$$= \lambda_j Y_j + \sum_{i=1}^{n} a_i Y_i.$$

(5.5)

J is a block-diagonal matrix, $\delta J_{k+1} \equiv QR_{k+1}Q^{-1}$, and δY_{k+1} *has been expanded in terms of a spectrum of the eigenvectors of* R_0, with

$$|a_i| \ll |\lambda_j|, |a_{i+1}| \ll |a_i|.$$

(5.6)

The point is that, provided R_0 has been 'tuned' so that this condition is true, the first few terms in the spectrum of this iteration of the eigenstate will contain most of the essential information about δR_{k+1}. This appears quite similar to the detection of color in the retina, where three overlapping non-orthogonal 'eigenmodes' of response are sufficient to characterize a huge plethora of color sensation. Here, if such a spectral expansion is possible, a very small number of observed eigenmodes would suffice to permit identification of a vast range of changes, so that the rate-distortion constraints become quite modest. That is, there will not be much distortion in the reduction from paths in R-space to paths in A-space.

Reflection suggests that, if consciousness indeed has something like a tunable retina – crudely, if 'the eye of the mind' has a fovea – then appropriately chosen observable correlates of consciousness may, at a particular time and under particular circumstances, actually provide very good local characterization of conscious process. Large-scale global processes are, of course, another matter.

Detailed reconsideration of the base paradigm – the visual retina – seems of interest.

2. The visual system

The tunable retina is, in fact, quite an old idea, as is the information-theoretic approach, which Schawbe and Obermayer (2002) describe as follows:

> Adaptation is a widespread phenomenon in nervous systems, and it happens on multiple time-scales, i.e. the activity-dependent refinement cortical maps (weeks), perceptual learning (hours and days) or contrast adaptation (seconds) in the primary visual cortex. It is reasonable to hypothesize that the functional role of these adaptation mechanisms is to provide flexibility to function under varying external conditions. Using concepts from information theory the specific idea that neuronal codes constitute efficient representations of the sensory world has been formulated (Attneave, 1954; Barlow, 1959; Atick, 1992). Subsequently the adaptation processes were mainly viewed as a signature of an ongoing optimization of sensory systems to changing environments as characterized by their statistical properties, i.e. as an optimization of the information transfer between the ensemble of stimuli and the neuronal responses.

Brenner et al. (2000) put it thus:

> One of the major problems in processing the complex dynamic signals that occur in the natural environment is providing an efficient representation of these data. More than 40 years ago, Attneave (1954) and Barlow (1961) suggested that steps in the neural processing of information could be understood as solutions to this problem of efficient representation. This idea was later developed by many groups, especially in the context of the visual system. Efficient representation requires a matching of the coding strategy to the statistical structure of incoming signals...
>
> The mean light level, for example, changes by orders of magnitude as we leave a sunny region and enter a forest. Adaptation to mean light level ensures that our visual responses are matched to the average signal in real time, thus maintaining sensitivity to the fluctuations around this mean.

The mechanism of light level tuning for the visual retina involves a shift from a band pass Fourier spatial frequency filter at elevated levels of luminance, where noise is not a major concern and high frequency spatial data can be processed, to a low frequency pass spatial frequency filter at low luminance, a regime where quantum noise dominates. Here, large shapes, without color, become the objects of attention. As Atick (1992) shows elegantly, quantum noise considerations can predict visual retina spatial filter performance from first principles, without much parameter fitting.

The model of section 5.1, which focused on altering operator spectral properties to determine Rate Distortion behavior, is roughly analogous. Rate Distortion arguments, unlike Atick's (1992) energy functional-analog minimization, are independent of the particular distortion measure chosen, but in a complicated ϵ - δ sense, which we briefly explore below.

For internal retinas like the one we propose for consciousness (or, elsewhere, for immune cognition; Wallace, 2003), generalized noise is not likely to have a simple quantum structure, and optimizations may not be at all straightforward.

The argument is as follows: suppose that the version of the 'real world' to be perceived by the internal retina has a high dimensional, and extremely complicated, alphabet, which is projected by that retina onto a simpler – e.g. lower dimensional, alphabet, so that information will inevitably be lost. Suppose that the full (internal) world can be characterized by an information source **X** and its retinal projection by a simpler information source **Y** such that paths of signals generated by **X**, of the form $x = x_0, x_2, ..., x_n, ...$, are mapped by some many-to-one operator R onto paths y. Use any distortion metric $d(x, y) = d(x, Rx)$ which measures the average deviation of x from y.

The Rate Distortion Theorem states that for any chosen maximum average distortion such that $d(x, Rx) < \epsilon$ there is a maximum possible transmission rate δ such that if **X** is mapped by R onto **Y** at a transmission rate (i.e. channel capacity) $C < \delta$, then the average distortion will be less than ϵ. The mutual information between **X** and **Y** $= R$**X** provides the essential splitting criterion.

If the organism has much time, then the retina operator R can indeed remain fixed, and its rate of operation simply slowed down until the ϵ constraint on error is matched.

This is clearly not an option for animals who are hunted (or hunt) in the night. The rate of signal recognition becomes very important, hence the change from spatial band pass to low pass filtering, as a means of maintaining transmission rate at the expense of perceived detail: tunable coarse-graining. See the Appendix for a brief example of coarse graining.

If consciousness has, as we all believe, quintessential survival value, then spectral tuning of R to optimize both ϵ and δ under changing conditions becomes likewise a priority, but the constraints may not be simply defined by quantum noise, as in the visual retina, and the elegant calculation of Atick and Redlich (1990) is not sufficient. Indeed, like biological universality class tuning, there appear to be whole sets of monotonic relations between ϵ and δ which are subject to tuning.

There are further complications: tunable internal retina arguments can be inverted to produce global structures in much the same way local tangent spaces can be linked together by an atlas structure to create a larger-scale differential geometry (e.g. Sternberg, 1964). That is, following our development, the fovea of the mind's eye is, in fact, a local projection of a high order or complex alphabet information source onto a lower order, simpler alphabet, information source, done in a manner to locally optimize certain rate-distortion factors. An algebraic geometer at this point can invoke any number of globalization theorems to canonically construct a larger embedding manifold with very interesting properties.

Such larger structures, however, are not unique, and not at all likely to be simple. We examine in more detail the argument-by-abduction from differential geometry.

3. The torus and the sphere

The tunable retina atlas we have proposed for dual information sources of cognitive processes is taken in concept from differential geometry, and an example can help show where this approach is leading.

Consider the two-dimensional torus and sphere within three dimensional space. The sphere is most simply defined as the set of points a fixed distance from some given point of origin. The torus is a little more complicated: take a 2-square in three dimensional space. Roll it so the top meets the bottom, then stretch the resulting cylinder until the ends meet. More directly, identify top and bottom edges of the square, and then identify the left and right edges.

These are fundamentally different constructions: Any closed one dimensional loop on the surface of a sphere may be continuously shrunk to a point. This is not true for the torus, since a closed loop which rings the torus cannot be shrunk down to a point, but is limited to size of the torus itself.

On the other hand, both structures are two dimensional surfaces in three-space. At any point on either a sphere or a torus, a 'small enough' patch containing that point can be mapped exactly onto a two dimensional tangent plane tuned to that point, without doing violence to the essential difference between the surfaces. This is analogous to our elementary tunable retina construction which locally maps a path of operators having m^2 components each onto a path of vectors having only m components, with minimal loss of information and maximal transmission rate.

The analogy with differential geometry is limited at best. We are quite definitely not proposing a pseudoriemannian geometry based on the 'Fisher-information metric'. Rather, the underlying manifold is an information source producing complex symbolic strings, and the R-projection onto a lower dimensional or coarse-grained information source is done by means of a local tuning which jointly minimizes distortion and maximizes transmission, subject to some embedding constraint structure defining the relation between them, which may itself be tunable. Distortion can be measured by any number of appropriate measures, according to the Rate Distortion Theorem. The result is far more like a stochastic version of a Finsler, rather than a Riemannian, system.

Such structure might well be called a Rate Distortion manifold, instantiated through a locally-tuned coarse graining which has a larger scale Rate Distortion topology.

Thus, while the tunable retina is postulated to be a local coarse-grained construction for cognitive processes like the tangent space in differential geometry, so that the 'redness of red' may well be empirically indistinguishable between individuals having normal vision, the threshold at which the red signal becomes conscious, its meaning once it becomes conscious, and the possible and likely responses of the individual to it, are conditioned by larger global - i.e. topological - structures. These structures reflect the interaction of constraints of individual development and learning with the embedding culture which conditions them, matters which will determine larger global properties. These are the cognitive and conscious analogs of the difference between the torus and the sphere.

It is possible to make these considerations explicit.

4. Expanding the workspace

The Rate Distortion and Joint Asymptotic Equipartition Theorems are generalizations of the Shannon-McMillan Theorem which examine the interaction of two information sources, with and without the constraint of a fixed average distortion or some particular transmission rate target. We conduct one more iteration, and require a generalization of the SMT in terms of the splitting criterion for triplets as opposed to single or double stranded patterns. The tool for this is at the core of what is termed *network information theory* (Cover and Thomas, 1991, Theorem 14.2.3).

Suppose there are three (piecewise adiabatically memoryless) ergodic information sources, Y_1, Y_2 and Y_3. Assume Y_3 constitutes a critical embedding context for Y_1 and Y_2 so that, given three sequences of length n, the probability of a particular triplet of sequences is determined by *conditional probabilities with respect to Y_3*:

$$P(Y_1 = y_1, Y_2 = y_2, Y_3 = y_3) =$$

$$\Pi_{i=1}^{n} p(y_{1i}|y_{3i}) p(y_{2i}|y_{3i}) p(y_{3i}).$$

(5.7)

That is, Y_1 and Y_2 are, in some measure, driven by their interaction with Y_3.

Then, in analogy with previous analyses, triplets of sequences can be divided by a splitting criterion into two sets, having high and low probabilities respectively. For large n the number of triplet sequences in the high probability set will be determined by the relation (Cover and Thomas, 1992, p. 387)

$$N(n) \propto \exp[nI(Y_1; Y_2|Y_3)],$$

(5.8)

where splitting criterion is given by

$$I(Y_1; Y_2|Y_3) \equiv$$

$$H(Y_3) + H(Y_1|Y_3) + H(Y_2|Y_3) - H(Y_1, Y_2, Y_3).$$

We can then examine mixed cognitive/adaptive phase transitions analogous to learning plateaus (Wallace, 2002b) in the splitting criterion $I(Y_1, Y_2|Y_3)$, which characterizes the synergistic interaction between Y_3, taken as an embedding context, and the cognitive processes characterized by Y_1 and Y_2. Again, the results are similar to the Gould-Eldredge model of evolutionary punctuated equilibrium.

Clearly the model can be expanded to any number of interacting information sources, $Y_1, Y_2, ..., Y_s$ conditional on an external context Z in terms of a splitting criterion defined by

$$I(Y_1; ...; Y_s|Z) = H(Z) + \sum_{j=1}^{s} H(Y_j|Z) - H(Y_1, ..., Y_s, Z),$$

(5.9)

where the conditional Shannon uncertainties $H(Y_j|Z)$ are determined by the appropriate direct and conditional probabilities.

This simple-seeming extension opens another Pandora's box in the study of 'mind-body interaction' and the impacts of culture and history on individual cognition. It provides a new tool for examining the interpenetration of a broad range of cognitive physiological, psychological, and social submodules – not just neural substructures – with each other and with embedding contextual cultural language so characteristic of human hypersociality, all within the further context of structured psychosocial stress. Chapter 6 explores the implications for understanding comorbid mind/body dysfunction, and provides a laundry list of physiological, psychological, and social cognitive modules associated with health and disease.

Bennett and Hacker (2003) define the 'mereological fallacy' in neuroscience as the assignment, to parts of an animal – here the brain – of those characteristics which are properties of the whole. Humans, through both their embedding in cognitive social networks, and their secondary epigenetic inheritance system of culture, are even more than 'simply' individual animals. Equation (5.9) implies the possibility of extending the global neuronal workspace model of consciousness to include both internal cognitive physiological systems and embedding cognitive and other structures, providing a natural approach to evading that fallacy.

Equation (5.9) is itself subject to significant generalization. The single information source Z is seen here as invariant, not affected by, but affecting, cross talk with the information sources for which it serves as the driving context. Suppose there is an interacting system of contexts, acting more slowly than the global neuronal workspace, but communicating within itself. It should be possible, at first order, to divide the full system into two sections, one 'fast,' containing the Y_j, and the other 'slow,' containing the series of information sources Z_k. The fast system instantiates the conscious neuronal workspace, including crosstalk, while the slow system constitutes an embedding context for the fast, but one which engages in its own pattern of crosstalk. Then the extended splitting criterion, which we write as

$$I(Y_1, ..., Y_j | Z_1, ..., Z_k),$$

(5.10)

becomes something far more complicated than equation (5.9). This must be expressed in terms of sums of appropriate Shannon uncertainties, a complex task which will be individually contingent on the particular forms of context and their interrelations.

This approach, while arguably more general than dynamic systems theory, can incorporate a subset of dynamic systems models through an appropriate coarse graining, a standard construction described at more length in Appendix A.

Again, the essential trick is to show that a system has a 'high frequency limit' so that an appropriate coarse graining catches the dynamics of fundamental importance, while filtering out 'high frequency noise'.

Taking this analysis into consideration, the model of equation (5.10) constitutes a 'double coarse-graining' in which the Z_k represent a 'slow' system which serves as a driving conditional context for the 'fast' Y_j of the global neuronal workspace.

It is possible to envision a 'multi' (or even distributed) coarse graining in which, for example, low, medium, and high, frequency phenomena can affect each other. The mathematics of such extension appears straightforward but is exponentially complicated. In essence one must give meaning to the notation

$$I(Y_1, ..., Y_j | X_1, ..., X_k | Z_1, ..., Z_q),$$

(5.11)

where the Y_j represent the fast-acting cognitive modules of the global neuronal workspace, the X_k are intermediate rate effects such as emotional structure, long-term goals, immune and local social network function, and the like, and the Z_q are even slower-changing factors such as cultural structure, embedding patterns of psychosocial stress, the legacy of personal developmental and community history, and so on.

Such analysis is consistent with, but clearly extends, the 'standard model' of global neuronal workspace theory.

Ultimately, culture, developmental history, and structured stress serve as essential contexts-of-context, in the sense of Baars and Franklin (2003), defining a further hierarchy of externally-imposed constraints to the functioning of individual consciousness. Equation 5.11 suggests a means of explicitly modeling those constraints.

5.   Energy efficiency and consciousness

5.1   Simple neural modules

A pioneering study by Levy and Baxter (1996) explores the energy costs of neural coding strategies, a matter which will prove to be of some interest here. To paraphrase Laughlin and Sejnowski (2003), detailed analysis comparing the representational capacity of signals distributed across a population of neurons with the costs involved suggests sparse coding schemes, in which a small proportion of cells signal at any one time, use little energy for signaling but have a high representational capacity because there are many different ways in which a small number of signals can be distributed among a large number of neurons. This is mitigated by the energetic cost of maintaining a large number of neurons, if they rarely signal. Thus there is an optimum proportion of active cells which depends on the ratio between the cost of maintaining a neuron at rest and the extra cost of sending a signal. When signals are relatively expensive, it is best to distribute a few of them among a large number of cells. When cells are expensive, it is more efficient to use few of them and to get all of them signaling.

A simplified version of the Levy and Baxter argument is as follows:

Suppose there are n binary neurons, taking an active value of 1 with probability p and an inactive value of 0 with probability $1-p, 0 \leq p \leq 1$. Classically, each binary neuron has a 'channel capacity' given by

$$h(p) = -p\log(p) - (1-p)\log(1-p).$$

(5.12)

See Ash (1990) or Cover and Thomas (1991) for details.

The maximum possible channel capacity of n such neurons would be the sum of n independent channels, so that

$$H(p) \leq nh(p).$$

(5.13)

If an active neuron has r times the energy requirements of an inactive one, then the average energy consumed by an active fraction p of n total neurons is just

$$E(p, r) = npr + n(1-p) = n(1 + p(r-1)),$$

(5.14)

where, again, energy units are measured in terms of an inactive neuron.

The ratio of maximum possible channel capacity to energy consumption is, then,

$$H(p)/E(p,r) \leq f(p,r) \equiv h(p)/(1 + p(r-1)),$$

(5.15)

independent of n.

Taking a typical value for r, say $r = 100$, so that a working binary neuron consumes 100 times the energy of a resting one (e.g. Lennie, 2003), then numerically solving the extremum problem

$$df(p,100)/dp = 0$$

for p gives $p = p^* \approx 0.0334$, so that the most energy efficient neural system, in this model, will have only about three percent of its neurons active at any one time, a startling 'sparse code' result.

The peak of $f(p,r)$ as a function of p for even large fixed r is actually very broad, having a significant full width at half maximum (FWHM). In the example for $f(p,100)$, half-maximum is met at $p = 0.0034, 0.3864$, so that FWHM= 0.3830, which is not inconsiderable.

Levy and Baxter (1996) examine a more complicated model which has, comparatively, a narrower peak than the simple binary neuron, but this too has a rather large FWHM, suggesting that the maximization of efficiency is at best highly approximate: large fractions of neurons may, apparently, be mobilized for short times, dependent on the ability to meet the energy demand.

Balasubramanian et al. (2001) examine more general metrics of neural efficiency, and the role of noise, using both a Lagrange multiplier and a complicated iterative Arimoto-Blahut optimization strategy, making empirical application to the distribution of burst sizes in the visual retina.

5.2   Interacting modules: the global workspace

Alternative – and perhaps competing – metrics may apply at higher levels of organization, in particular to the global neuronal workspace itself.

Figures 1 and 2 show the results of a simple calculation at a level more complicated than that represented by equations (5.13) and (5.14). Suppose that neural modules interact within themselves according to what might be characterized as 'strong ties', i.e. those which disjointly partition a structure according

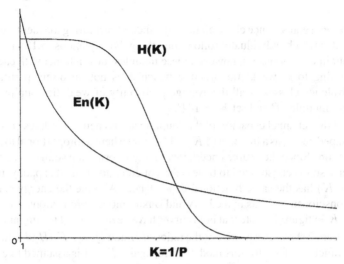

Figure 1. Channel capacity-source uncertainty and energy consumption of a system of interacting neural modules as a function of the inverse probability of weak ties coupling the modules, $K = 1/P$. H is taken as proportional to an error function in K, and energy consumption as a linear function in P, i.e. Inverse in K.

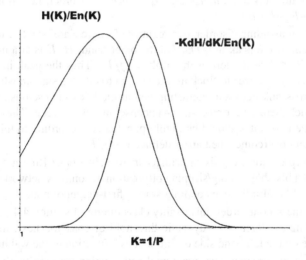

Figure 2. Source uncertainty per unit energy and disorder $Q = [-K dH/dK]$ per unit energy, according to the model of Figure 1. We assume that, for 'social' systems, cognitive energy efficiency is to be maximized, while the experience of disorder is to be minimized, which are competing requirements. Clearly, two stable regimes are possible to this model: conscious, or unconscious/sleeping, depending on the relative weighting of the optimization, which may, indeed, change according to resource availability: when tired, one falls asleep.

to some equivalence class relationship, indeed permitting the identification of modules which individually follow something like equations (5.13) and (5.14). To attain consciousness, however, these modules must interact with each other according to a 'weak' tie structure which does not disjointly partition into equivalence classes. Call the average probability of weak tie coupling across neural modules P and set $K = 1/P$.

Let the 'channel capacity' of the coupling across neural modules be a reverse S-shaped curve, as a function of $K = 1/P$, taken here as proportional to an Error Function. Since the source uncertainty of any information source - cognitive or otherwise - is constrained to be less than or equal to channel capacity, in figure 1, $H(K)$ has the same functional dependence. Assume the energy consumed by consciousness is, as in the Levy and Baxter model, proportional to P, hence to $1/K$ in figure 1. Note that both curves have been adjusted to similar maxima.

Figure 2 shows, respectively, the ratios of H and $Q = -KdH/dK$ to energy, as a function of K, for this model. Here, again, $P = 1$ is assumed to consume 100 times the energy of $P = 0$. The plots have also been adjusted to similar maxima.

Again, for a complex 'social' construction, as opposed to the individual elementary structures which compose it, while attempting to maximize H/E, the system attempts to *minimize* the experience of disorder, i.e. to minimize $Q/E = [-KdH/dK]/E$.

In terms of avoiding disorder, two regimes of figure 2 are 'stable': to the left and the right of the peak in Q/E. On the other hand, H/E is at a maximum for some $P < 1$ to the left of the peak in Q/E. Thus the peak in $Q/E = [-KdH/dK]/E$ serves to 'lock in' the system to either a state in which there is much cross-talk between interacting neural modules - consciousness - or a state in which there is little cross-talk – unconsciousness or sleep. The transition between the two states should be highly punctuated, according to this model, as the system overcomes the barrier defined by Q/E.

This interpretation is consistent with recent work by Lopez-Ruiz et al. (2004), who found a bistable waking/sleeping bifurcation in complex networks defined by mean-field multiplicative coupling among first-neighbor nodes.

Clearly the second order universality class tuning of section 4.4, which we use to define the fluctuating dynamic threshold of consciousness, makes the most sense on the left hand side of figure 2, as a function of the waking state.

Questions of energy use vs. functional optimization for cognitive/conscious processes require further study, particularly as regards the impacts of multiple parallel or hierarchical organization levels. The difference between these two examples – the efficiency of neural modules vs. that of assemblies of interacting cognitive modules – might be seen as analogous to the contrast between the interest of individuals within an organization, who may wish to optimize their personal income per unit effort, vs. the interest of the organization itself, which is attempting to maximize its 'market share'. These are not at all the same goals, and the organizational priorities will likely be at considerable odds with the individual interests of the employees. Under a hierarchy, then, optimization may become a matter of conflict between competing levels. The FWHM of the different optimizations may represent the solution to mitigating that conflict.

As the calculation at the end of Chapter 4 indicates, these matters may rapidly become quite complicated mathematically.

5.3   Reconsidering fMRI

Recently Shulman et al. (2003) reexamined functional magnetic resonance imaging (fMRI) from the perspective of global workspace theory. They find that the high energy consumption of the brain at rest and its quantitative usage for neurotransmission reflect a high level of neuronal activity for the non-stimulated brain. This high activity, in their view, supports a reinterpretation of functional imaging data; e.g., where the large baseline signal has commonly been discarded. Independent measurements of energy consumption obtained from calibrated fMRI equaled percentage changes in neuronal spiking rate measured by electrodes during sensory stimulation at two depths of anesthesia. These quantitative biophysical relationships between energy consumption and neuronal activity, they claim, provide novel insights into the nature of brain function. They propose the high resting brain activity includes the global interactions constituting the subjective aspects of consciousness. Anesthesia, by lowering the total firing rates, correlates with the loss of consciousness. Shulman et al. conclude that these results, which measure localized neuronal response and distinguish inputs of peripheral neurons from inputs of neurons from other brain regions, fit comfortably into the global neuronal workspace model of Dehaene, Changeux, and others.

One is tempted to interpret the 'large baseline signal' in terms of the operator R_0 of equation (5.3), and the fMRI-measured differences in terms of that equation's δR_{k+1}, leading 'naturally' into the tunable retina arguments associated with the Rate Distortion manifold/atlas topologies discussed above.

A second temptation is to reformulate the parametization of the consciousness phase transition itself in terms of baseline energy consumption, possibly producing a series of punctuated changes corresponding to different levels of consciousness, much as shown in figures 1 and 2.

These questions remain to be studied.

6.   Quantum systems

6.1   Quantum neural networks and information theory

A recent spate of publications suggests that microtubule structures within neuronal cells might be the site of an exotic physics making possible the classical-scale operation of a brain-wide quantum coherence producing consciousness (Hameroff, 2001; Penrose, 2001; Hameroff et al., 2002). Such speculations were scathingly analyzed by Tegmark (2000), who, based on a calculation of neural decoherence rates, argued that the degrees of freedom of the human brain that relate to cognitive processes should be thought of as a classical rather than a quantum system. There is, in Tegmark's view, nothing fundamentally wrong with the current classical approach to, for example, neural network simulation. Tegmark found that the decoherence time scales, $10^{-13} - 10^{-20}$ sec., are typically much shorter than the relevant dynamical time scales ($\approx 10^{-3} - 10^{-1}$ sec.), both for regular neuron firing and for kinklike polarization excitations in microtubules. Tegmark's result thus disagrees with suggestions by Penrose and others that the brain acts as a quantum computer, and that quantum coherence is related to consciousness in a fundamental way.

Suppose, however, a large quantum neural network is held at a few micro or mille degrees Kelvin, $10^{-6} - 10^{-3}$ degrees Centigrade above absolute zero. Quantum decoherence times typically scale exponentially with temperature, and, cold enough (and/or with sufficient error correction machinery), dynamical time scales in Tegmark's range of $10^{-3} - 10^{-1}$ sec. might well be possible across fairly large structures. Could these, then, become conscious, in the sense of the formal treatment of global neuronal workspace theory given here?

The answer does not seem entirely straightforward, since the analysis depends on a homology between the information source uncertainty dual to a cognitive process and free energy density on the one hand, and on a further, i.e. second order, extension of that treatment to universality class tuning of phase transitions associated with the first process – some version of a fluctuating dynamic threshold. This is a difficult enough sequence in a purely classical structure, and the information dynamics of quantum systems are not particularly well characterized: In spite of much physics literature *Sturm und Drang*, there are few mathematically rigorous quantum generalizations of the fundamental results of information theory.

One recent attempt was the quantum extension of the Shannon-McMillan Theorem by King and Lesniewski (1998) (KL), which serves as an interesting starting point. We paraphrase results from that paper, recapitulating as well as some standard material on quantum information sources.

The fundamental object of interest in KL is a quantum system whose state space is a tensor product of many copies of one fundamental space M. The source produces a signal which is encoded by a state in M, and the ensemble of possible states is represented by a density operator ρ on M. An extended source corresponds to a sequence of such states, which is interpreted as a message. The probabilistic nature of the message is contained in the density operator on the tensor product of copies of M. If that operator is the product $\rho \otimes ... \otimes \rho$, there are no correlations between signals in the message, the 'quantum Bernoulli' source which was the focus of Schumacher's work (Schumacher, 1995, 1996). A more useful quantum information source is one in which there are correlations on all time scales between signals in the message. The density operator then becomes a much more complex object.

For purely classical signals, the Shannon-McMillan Theorem permits the splitting of all possible signals into two classes, a relatively small number of 'meaningful' ones with significant probabilities, and a much larger number with vanishingly small probability. The criterion for splitting is the uncertainty of the information source. The quantum result is a splitting of the state space into relevant and irrelevant subspaces, with the Von Neumann entropy as the criterion.

KL derive an estimate for the dimension of the relevant subspace by computing the entropy of a classical source obtained by taking measurements on the quantum system. For the case of a quantum source emitting orthogonal states, the Von Neumann entropy is the same as the uncertainty of the associated classical information source, since the density operators all commute, and the Shannon-McMillan Theorem is recovered exactly. Non-orthogonal sources are more complicated, and their result only provides somewhat loose limits on the dimensionality of the relevant space.

The quantum source sends a series of signals, each of which is a vector in a finite dimensional Hilbert space $\mathcal{H}$. The source is taken as discrete, with each signal an element of a finite set $\mathcal{S} = |\psi_1 >, ..., |\psi_s >$ of normalized vectors in $\mathcal{H}$. We take $\mathcal{H}$ as spanned by $\mathcal{S}$, so that $\mathcal{H}$ is of dimension $d \leq s$. Unlike the classical case, this system can entertain a superposition of states. Let p_j be the given probability of the state $|\psi_j >$ being sent. The density matrix corresponding to the ensemble of signals $\mathcal{S}$ is then

$$\rho = \sum_{1 \leq j \leq s} p_j |\psi_j >< \psi_j|,$$

(5.16)

with $tr(\rho) \equiv 1$, where tr is the trace operator. While $\mathcal{S}$ and the distribution of p_j uniquely determine the density matrix, each such matrix corresponds to an infinite number of possible sets of states.

The observables associated with quantum signals are $d \times d$ hermitian matrices, the elements of a C^*-algebra $\mathcal{A} = \mathcal{L}(\mathcal{H})$ of linear observables on $\mathcal{H}$. The state on the algebra of observables $\mathcal{A}$ associated with the density matrix ρ is, for any given $A \in \mathcal{A}$,

$$\tau_1(A) \equiv tr(A\rho) = \sum_{1 \leq j \leq s} p_j < \psi_j|A|\psi_j > .$$

(5.17)

Appropriate generalizations can be given for infinite dimensional tensor products, and ergodic quantum information sources can be defined.

The density matrix of order n becomes, in terms of the states ψ_j which span $\mathcal{S}$,

$$\Pi_n = \sum_{1 \leq j_1,\ldots,j_n \leq s} p_{j_1,\ldots,j_n} |\psi_{j_1} >< \psi_{j_1}| \otimes \ldots \otimes |\psi_{j_n} >< \psi_{j_n}|.$$

(5.18)

The entropy associated with a sequence of n signals is defined as

$$H_n(\Pi) \equiv -tr_{\mathcal{H}^{\otimes n}}(\Pi_n \log \Pi_n).$$

(5.19)

Some development gives

$$H_{m+n}(\Pi) \leq H_m(\Pi) + H_n(\Pi),$$

so that the limit

$$h(\Pi) = \lim_{n \to \infty} \frac{H_n(\Pi)}{n}$$

(5.20)

exists. We call $h(\Pi)$ the entropy of the quantum source. For a Bernoulli source $\Pi_n = \rho \otimes ... \otimes \rho$ and $h(\Pi) = -tr_{\mathcal{H}}(\rho \log \rho)$. General sources with internal serial correlations have far more complex expressions for h.

Let $\mathbf{A} = [A_1, ..., A_r], r < \infty$ be a family of observables on $\mathcal{H}$ such that $A_j \geq 0$ for all j, and

$$A_1 + ... + A_r = I,$$

(5.21)

where I is the identity. We call the set $\chi_{\mathbf{A}} = [1, ..., r]$ the classical alphabet associated with $\mathbf{A}$, and denote by $\chi_{\mathbf{A}}^{\infty}$ the space of all infinite messages over the alphabet $\chi_{\mathbf{A}}$. In this way we can associate a classical information source with each quantum information source.

Let $\mathcal{H}^{\otimes n}$ be the space of all signals of length n for an ergodic quantum information source. According to the KL version of the quantum Shannon-McMillan Theorem, it can be factored into two orthogonal subspaces

$$\mathcal{H}^{\otimes n} = \mathcal{S}_n \otimes \mathcal{S}_n^{\perp},$$

(5.22)

whose relative dimensions are constrained by the uncertainty of the classical information source $h_{\mathbf{A}}$ associated with the quantum source in a precise manner. If the $|\psi_j >$ are orthogonal, $h_{\mathbf{A}}$ is just the Von Neumann entropy of the source, since the density operators all commute.

Let $P_{\mathcal{S}_n}$ be the orthogonal projection onto the relatively small subspace $\mathcal{S}_n$.

Let C be an observable $C \in \mathcal{L}(\mathcal{H}^{\otimes n})$, where the signal is of length n. Then, according to the KL form of the quantum Shannon-McMillan Theorem, the difference

$$|\tau(CP_{\mathcal{S}_n}) - \tau(C)|$$

can be made arbitrarily small as n increases without limit. Here τ is an appropriate infinite-dimensional generalization of τ_1 above, in terms of the complicated density matrices Π.

Again paraphrasing KL, in the case where the $|\psi_j>$ are orthogonal, there is a direct correspondence with the classical Shannon-McMillan theorem, and the quantum theory is simply a restatement of the classical result, with the associated classical source uncertainty h_A, (which constrains the dimensionality of the significant space $\mathcal{S}_n$), given by the Von Neumann entropy.

The KL version of the quantum SMT is deficient in that it does not provide an exact parallel to the asymptotic equipartition property (AEP), and KL's relevant subspace can be shown not to be minimal in general, although it is so for the Bernoulli case. This is discussed in the more recent work of Bjelakovic et al. (2003, 2004), who do, in fact establish the quantum AEP under very general conditions. They proceed, following the work of Hiai and Petz (1994), from the fact that a classical subsystem – the maximal abelian subalgebra of the entire non-commutative algebra of observables – has Shannon entropy equal to the von Neumann entropy of the full n-block quantum state.

Restricting the given quantum state to an appropriate classical lattice system produces a classical system related to the original system, and one can build further from this. The essential contribution of Bjelakovic et al. is to show that, only requiring simple ergodicity, the asymptotic limit on the dimension of a typical subspace is, in fact, for an n- block, given by $\exp[ns]$ where s is the mean von Neumann entropy. This gives an appropriate quantum SMT which addresses the failings of the KL approach and produces the asymptotic equipartition property which we find so useful in modeling the global neuronal workspace.

Although there may not yet be fully adequate quantum forms of the Shannon Coding Theorem, or its 'Learning Theorem' and Rate Distortion variants, these considerations nonetheless suggest a possible correspondence principle generalization of the classical neural network results given in the earlier sections: Parametization of the quantum information source corresponding to a QNN must reflect the underlying structural hierarchy of the system, incorporated in the renormalization symmetry and other inherent properties of the information source. Measurement must give an appropriately parametized classical information source with appropriate renormalization and generalized Onsager properties.

The parametization of the quantum information source might well be complicated, for example simultaneously involving both quantized and unquantized physical quantities. One imagines simultaneously macroscopic external signals and an array of quantum oscillators coupled by some kind of quantized field - phonons, photons, etc.

Since a quantum information source is still a 'language,' in the sense of the earlier sections of this work, its renormalization and generalized Onsager properties may not be simple extensions or reflections of commonly understood physical systems, but characterize, in no small part, the patterns of internal correlations defining that language – the jointly defined grammar and syntax of the coupling of sensory signal, neural weights and array of nonlinear oscillators constituting the system: Neural networks, quantum or classical, are defined by their 'meaning' even more than by their physical structure.

Some version of the various 'natural' relations between network architecture, learning paradigms, renormalization symmetry and generalized Onsager relations which applied to classical systems would seem appropriate to the pure quantum case as well. These extensions will not likely much resemble the classical case, however, no more than the quantum hydrogen atom resembles a planetary system, or superfluid helium resembles liquid water, a matter we will return to below.

6.2   Density matrix and path integral

Rojdestvenski and Cottam (2000), in their application of the results of Wallace and Wallace (1998) to physical processes, conclude with the following observation:

> If one takes an 'evolution' equation of any system..., it may always be written in the following differential form
>
> $$\psi(t + dt) = (1 + \mathbf{E}dt)\psi(t),$$
>
> where $\mathbf{E}$ is called the 'evolution operator.' If the evolution has different 'channels,' i.e.
>
> $$\mathbf{E} = \sum_{i=1}^{N_0} \mathbf{E}_i,$$
>
> then [the first equation] takes the following recursive form:
>
> $$\psi(t + mdt) = (1 + \mathbf{E}dt(...(1 + \mathbf{E}dt(1 + \mathbf{E}dt(1 + \mathbf{E}dt)))...)\psi(t) =$$
>
> $$\sum_{r=1}^{m}(dt)^m \sum_{C_r} K(C_r)[\mathbf{E}_{i_1}...\mathbf{E}_{i_r}]\psi(t),$$
>
> and again we deal with the 'sentence' representation. In a certain sense, any temporal evolution, if only it is describable by equations, is a message [from some information source] in its own right.

Behrman et al. (1996) open their description of a quantum dot neural network in a similar manner:

In most artificial neural network implementations, the neurons receive inputs from other processors *via* weighted connections and calculate an output which is passed on to other neurons. The calculated output... of the ith neuron [is determined from] the signals from the other neurons in the network... Similarly we can write the expression for the time evolution of the quantum mechanical state of a system:

$$|\psi(x_f, T) >= G(x_f, T; x_0, 0)|\psi(x_0, 0) > ...$$

Here $|\psi(x_0, 0) >$ is the input state, the initial state of the quantum system. $|\psi(x_f, T) >$ is the output state, the state of the system at $t = T$. G is the Green's function, which propagates the system forward in time, from initial position x_0 at time $t = 0$ to final position x_f at time $t = T$. [G can be expressed] in the Feynman path integral formulation of quantum mechanics (Feynman, 1965), in which G is thought of as the infinite sum over all possible paths that the system could possibly take to get from x_0 to x_f... Each path is weighted by the complex exponential of the phase contributed by that path, given by the classical action for that path;... Each of the N [quantum] neurons' different possible states contribute to the final measured state; the amount it contributes can be adjusted by changing the potential energy...

Those paths with higher weighting thus have higher probability (and are 'meaningful,' in our terminology), than the others. For an 'ergodic' information source such paths would be equiprobable.

Using this formalism, Behrman et al. (1996) conclude that

Potentially, a quantum neural network would be an extremely powerful computational tool... capable, at least in principle, of performing computations that cannot be done, classically... an actual working quantum neural net would likely want to take advantage of the greater multiplicity and connectivity inherent in an entire array of quantum dot molecules, by placing molecules physically close enough to each other that nearest neighbors can interact directly...

The path integral formulation of quantum density matrices (Feynman, 1998) thus seems to constitute the natural linkage between quantum mechanics and quantum information theory in much the same way that the Large Deviations Program of applied probability connects statistical mechanics, fluctuations and information theory in classical systems. Imposition of appropriate renormalization symmetry on the ergodic quantum information source dual to the QNN, in the context of a similarly appropriate 'generalized Onsager relation' and associated algebras, would indeed seem to be the most natural means of expressing the unique architecture of the network, hierarchical or otherwise.

By analogy, it seems that a Landau-like 'two fluid' model of superconductivity and superfluidity might apply to the general QNN, with a classical information source uncertainty playing the role of a 'phonon gas excitation' of the purely quantum QNN (Feynman, 1998).

6.3 Speculations

The development of King and Lesniewski (1998), successfully generalized by Bjelakovic et al. (2003, 2004), is an attempt to rigorously extend the

Shannon-McMillan Theorem to quantum systems. In conjunction with the material described elsewhere in this work, the approach suggests a direction for development of a purely quantum neural network formalism, in contrast, for example, with the quasi-classical results of Toth et al. (1996). Quantum neural networks, like their classical counterparts, should be reducible to the convolution of external 'sensory' activity, internal ongoing activity 'neural weights' and an array of nonlinear components into a single quantum information source parametized by continuous or quantized variates. 'Tuning' the parameters and the 'ongoing activity' should, as for classical systems, result in highly efficient pattern recognition, depending on the inherent grammar and syntax of the associated quantum information source: data consistent with the system's linguistic rules are recognized and acted on, others are not. The inherently parallel nature of pure quantum computation should provide some significant advantages over classical neural network pattern recognition. Quantum neural architecture should, as in the classical case, express itself in the renormalization symmetry of the dual quantum information source, its 'generalized Onsager relations,' and the algebraic structure of the underlying state space. Thus, for a certain class of QNN, high probability paths will define a quantum information source having grammar, syntax and higher order structures which will define the characteristics of the system for pattern recognition.

A kind of quantum linguistics – the extended algebra of Π operators corresponding to quantized neural networks – seems likely to be of some considerable interest.

Rigorous extension of such a theory to the second order tunable universality class effects needed for operation of a global neuronal workspace – the basis for consciousness – may not, however, be at all straightforward for quantum systems. In fact, if such extension is indeed possible, the kind of consciousness available to quantum structures is highly unlikely to at all resemble that of classical systems, putting yet another nail in the coffin of the Penrose treatment. This is because the differences with the classical case may far transcend the question of a half-second quantum coherence time: Even 'rapid' quantum consciousness may diverge appreciably from the classical variety with which we are all so intimately familiar, much as most quantum systems differ appreciably from their classical analogs.

The picture which comes to mind is the difference between a cryogenic flask of superfluid helium and a glass of water. The latter is often of considerable utility, while the former is usually only of academic interest.

More precisely, consideration suggests that, given an underlying QNN instantiating a quantum cognitive process, the information source associated with a second-order fluctuating dynamic threshold defined by universality class tuning might well itself be classical or semi-classical, producing a theory roughly analogous to what we have described here, although having a quantum base.

A far more challenging circumstance would arise if that second order tuning were itself associated with a fully quantum information source, for which the theory we have presented is wholly insufficient. It is a matter of conjecture, at this point, whether any such development is indeed possible. In the event it is, one might well be inclined to some quite odd astrobiological and other speculations.

Among these is the thought that conscious quantum neural networks engaged in mission-critical tasks are likely to have very strange failure modes, analogous, perhaps, to exotic mental disorders in humans. This problem may, in fact, impose a serious limit on the practical utility of most forms of machine consciousness, which have not had the benefit of several hundred million years of variation and selection.

Chapter 6

WHERE DOES ALL THIS LEAD?

1. Sociocultural context as selection pressure

The hierarchy of contexts is clearly able to write an image of itself on the function of the global neuronal workspace of consciousness, but what, really, are the mechanisms which instantiate equation (5.11)? A recent paper examines the similarities and contrasts between 'learning plateaus' in neural networks and punctuated equilibrium in evolutionary process from a perspective similar to that presented here (Wallace, 2002b). The starting point for that discussion was the information theory approach of Ademi et al. (2000) to evolutionary process in which they concluded that genomic complexity can be identified with the amount of information a gene sequence stores about its environment. This storage often occurs in a punctuated manner analogous to a learning plateau in a neural structure (e.g. Gould and Eldredge, 1977), although evolution is not at all a cognitive process. Cognition requires the active selection of one out of a complex repertoire of possible responses to a sensory or other input, based on comparison with a learned internal representation of the world. While genes do indeed constitute a kind of memory of past interaction with the world, response to selection pressure is not through direct comparison with that memory, but rather through the reproductive success of a random variation constrained by the path of evolutionary history.

This is not cognition, and there is no 'intelligent purpose' to adaptive or evolutionary process per se. Nonetheless, selection pressures are most often systematic patterns of interaction with an embedding and highly structured ecosystem in which each species is itself manifest through interpenetration (Lewontin 2000). Those ecosystems, acting as selection pressures, write images of themselves on gene sequences through reproductive success.

The slowly-acting factors of sociocultural structure and history in equation (5.11) – the Z_q – appear closely akin to selection pressures in the manner they write images of themselves on the dynamic global neuronal workspace, the fast-acting, linked, cognitive neural modules represented by the Y_j.

That is, given the Z_q, those individuals who are able to limit their consciousness, or at least their consciously-driven behaviors, to the selection filter defined by the embedding sociocultural constraints, will be successful in their life courses. That very success will, according to this perspective, virtually sculpt their individual conscious lives, limiting – indeed, defining – both what can be perceived and experienced and what can be carried out as voluntary activity.

This effect, which should be very strong, seems largely masked by a highly efficient form of perceptual completion.

Section 5.2 explored the widespread nature of adaptation, the refinement of cortical maps, perceptual learning, and, in particular, contrast adaptation in the retina. A similarly common phenomenon is perceptual completion, the neural filling-in of 'holes' in perception. Blind-spots in the visual field seem to disappear, rapid sequential blink is taken to be continuous motion, and the like (e.g. Welchman and Harris, 2003; Zur and Ullman, 2004; Lerner, Harel, and Malach, 2004). Arguing by abduction from these examples, it seems likely that 'the fovea of the mind's eye' – consciousness as we have interpreted it, involving a second order retina-like adaptive structure – may also engage in perceptual completion. Thus the sociocultural sculpting of individual conscious life may not be easily recognized by an individual. Not that consciousness itself is an illusion, but rather the apparent seamless continuity of the 'I-of-the-hurricane' centrality of conscious experience and voluntary action is, in fact, largely a projective construct, filling in around the holes inevitably left by sociocultural conditioning and the path dependent outcome of individual development and history.

The mathematics which intrudes here is precisely the tangent construction of the sphere/torus example. Restricting study to the tangent plane at a point on a torus, one cannot find the central hole, which is a higher order structure. Indeed, points on the torus are all equally accessible to each other through paths restricted to the torus surface. The dance of consciousness seems uninterrupted to the dancer, yet whole realms of possibility are structurally inaccessible, a fact which can only be determined by relatively sophisticated analysis, the analog of trying to shrink loops on a torus to a point. We begin to require an understanding of the large-scale topological properties of rate distortion manifolds.

These considerations have particular, and very disturbing, implications regarding the social induction of disorders of consciousness and cognitive process such as psychopathy and sociopathy (Mealey, 1995), a matter explored at greater length below. We can, in fact, place discussions about the social induction of pathology in a far more comprehensive context.

2. Autocognitive developmental disorder

Consciousness is not the only cognitive physiological or psychological phenomenon, and much of what we have done can be applied to other – interacting – cognitive modules of interest. We begin with a brief listing of a few of them, and then explore how structured psychosocial stress can impose an image of itself upon human development, entraining not only consciousness, but other cognitive modules into characteristic patterns of mind/body dysfunction in which disorders of consciousness are part of a larger whole.

2.1 Immune function

Atlan and Cohen (1998) have proposed an information-theoretic cognitive model of immune function and process, a paradigm incorporating cognitive pattern recognition-and-response behaviors analogous to those of the central nervous system. This work follows in a very long tradition of speculation on the cognitive properties of the immune system (Tauber, 1998; Podolsky and Tauber, 1998; Grossman, 1989, 1992, 1993a, b, 2000).

From the Atlan/Cohen perspective, the meaning of an antigen can be reduced to the type of response the antigen generates. That is, the meaning of an antigen is functionally defined by the response of the immune system. The meaning of an antigen to the system is discernible in the type of immune response produced, not merely whether or not the antigen is perceived by the receptor repertoire. Because the meaning is defined by the type of response there is indeed a response repertoire and not only a receptor repertoire.

To account for immune interpretation Cohen (1992, 2000) has reformulated the cognitive paradigm for the immune system. The immune system can respond to a given antigen in various ways. It has 'options'. Thus the particular response we observe is the outcome of internal processes of weighing and integrating information about the antigen.

In contrast to Burnet's view of the immune response as a simple reflex, it is seen to exercise cognition by the interpolation of a level of information processing between the antigen stimulus and the immune response. A cognitive immune system organizes the information borne by the antigen stimulus within a given context and creates a format suitable for internal processing; the antigen and its context are transcribed internally into the 'chemical language' of the immune system.

The cognitive paradigm suggests a language metaphor to describe immune communication by a string of chemical signals. This metaphor is apt because the human and immune languages can be seen to manifest several similarities such as syntax and abstraction. Syntax, for example, enhances both linguistic and immune meaning.

Although individual words and even letters can have their own meanings, an unconnected subject or an unconnected predicate will tend to mean less than does the sentence generated by their connection.

The immune system creates a 'language' by linking two ontogenetically different classes of molecules in a syntactical fashion. One class of molecules are the T and B cell receptors for antigens. These molecules are not inherited, but are somatically generated in each individual. The other class of molecules responsible for internal information processing is encoded in the individual's germline.

Meaning, the chosen type of immune response, is the outcome of the concrete connection between the antigen subject and the germline predicate signals.

The transcription of the antigens into processed peptides embedded in a context of germline ancillary signals constitutes the functional 'language' of the immune system. Despite the logic of clonal selection, the immune system does not respond to antigens as they are, but to abstractions of antigens-in-context.

2.2   Tumor control

Reflection shows the next cognitive submodule after the immune system is necessarily a tumor control mechanism that may include 'immune surveillance', but certainly transcends it. Nunney (1999) has explored cancer occurrence as a function of animal size, suggesting that in larger animals, whose lifespan grows as about the 4/10 power of their cell count, prevention of cancer in rapidly proliferating tissues becomes more difficult in proportion to size. Cancer control requires the development of additional mechanisms and systems to address tumorigenesis as body size increases – a synergistic effect of cell number and organism longevity. Nunney concludes

> This pattern may represent a real barrier to the evolution of large, long-lived animals
> and predicts that those that do evolve ... have recruited additional controls [over those
> of smaller animals] to prevent cancer.

Different tissues may have evolved markedly different tumor control strategies. All of these, however, are likely to be energetically expensive, permeated with different complex signaling strategies, and subject to a multiplicity of reactions to signals, including those related to psychosocial stress. Forlenza and Baum (2000) explore the effects of stress on the full spectrum of tumor control, ranging from DNA damage and control, to apoptosis, immune surveillance, and mutation rate. Elsewhere (Wallace et al., 2003) we argue that this elaborate tumor control strategy, particularly in large animals, must be at least as cognitive as the immune system itself, which is one of its components: some comparison must be made with an internal picture of a 'healthy' cell, and a choice made as to response: none, attempt DNA repair, trigger programmed cell death, engage in full-blown immune attack. This is, from the Atlan/Cohen perspective, the essence of cognition.

2.3 The HPA axis

The hypothalamic-pituitary-adrenal (HPA) axis, the 'flight-or-fight' system, is cognitive in the Atlan/Cohen sense. Upon recognition of a new perturbation in the surrounding environment, memory and brain or emotional cognition evaluate and choose from several possible responses: no action needed, flight, fight, helplessness (flight or fight needed, but not possible). Upon appropriate conditioning, the HPA axis is able to accelerate the decision process, much as the immune system has a more efficient response to second pathogenic challenge once the initial infection has become encoded in immune memory. Certainly 'hyperreactivity' in the context of post-traumatic stress disorder (PTSD) is a well known example. Chronic HPA axis activation is deeply implicated in visceral obesity leading to diabetes and heart disease, via the leptin/cortisol diurnal cycle (Bjorntorp, 2001).

2.4 Blood pressure regulation

Rau and Elbert (2001) review much of the literature on blood pressure regulation, particularly the interaction between baroreceptor activation and central nervous function. We paraphrase something of their analysis. The essential point, of course, is that unregulated blood pressure would be quickly fatal in any animal with a circulatory system, a matter as physiologically fundamental as tumor control. Much work over the years has elucidated some of the mechanisms involved: increase in arterial blood pressure stimulates the arterial baroreceptors which in turn elicit the baroreceptor reflex, causing a reduction in cardiac output and in peripheral resistance, returning pressure to its original level. The reflex, however, is not actually this simple: it may be inhibited through peripheral processes, for example under conditions of high metabolic demand. In addition, higher brain structures modulate this reflex arc, for instance when threat is detected and fight or flight responses are being prepared. Thus blood pressure control cannot be a simple reflex. It is, rather, a broad and actively cognitive modular system which compares a set of incoming signals with an internal reference configuration, and then chooses an appropriate physiological level of blood pressure from a large repertory of possible levels – a cognitive process in the Atlan/Cohen sense. The baroreceptors and the baroreceptor reflex are, from this perspective, only one set of a complex array of components making up a larger and more comprehensive cognitive blood pressure regulatory module.

2.5   Emotion

Thayer and Lane (2000) summarize the case for what can be described as a cognitive emotional process. Emotions, in their view, are an integrative index of individual adjustment to changing environmental demands, an organismal response to an environmental event that allows rapid mobilization of multiple subsystems. Emotions are the moment-to-moment output of a continuous sequence of behavior, organized around biologically important functions. These 'lawful' sequences have been termed 'behavioral systems' by Timberlake (1994).

Emotions are self-regulatory responses that allow the efficient coordination of the organism for goal-directed behavior. Specific emotions imply specific eliciting stimuli, specific action tendencies (including selective attention to relevant stimuli), and specific reinforcers. When the system works properly, it allows for flexible adaptation of the organism to changing environmental demands, so that an emotional response represents a *selection* of an appropriate response and the inhibition of other less appropriate responses from a more or less broad behavioral repertoire of possible responses. Such 'choice' leads directly to something closely analogous to the Atlan and Cohen language metaphor.

Damasio (1998) concludes that emotion is the most complex expression of homeostatic regulatory systems. The results of emotion serve the purpose of survival even in nonminded organisms, operating along dimensions of approach or aversion, of appetition or withdrawal. Emotions protect the subject organism by avoiding predators or scaring them away, or by leading the organism to food and sex. Emotions often operate as a basic mechanism for making decisions without the labors of reason; that is, without resorting to deliberated considerations of facts, options, outcomes, and rules of logic. In humans learning can pair emotion with facts which describe the premises of a situation, the option taken relative to solving the problems inherent in a situation, and, perhaps most importantly, the outcomes of choosing a certain option, both immediately and in the future. The pairing of emotion and fact remains in memory in such a way that when the facts are considered in deliberate reasoning when a similar situation is revisited, the paired emotion or some aspect of it can be reactivated. The recall, according to Damasio, allows emotion to exert its pairwise qualification effect, either as a conscious signal or as nonconscious bias, or both. In both types of action the emotions and the machinery underlying them play an important regulatory role in the life of the organism. This higher order role for emotion is still related to the needs of survival, albeit less apparently.

Thayer and Friedman (2002) argue, from a dynamic systems perspective, that failure of what they term 'inhibitory processes' which, among other things, direct emotional responses to environmental signals, is an important aspect of psychological and other disorder. Sensitization and inhibition, they claim, 'sculpt' the behavior of an organism to meet changing environmental demands.

When these inhibitory processes are dysfunctional – choice fails – pathology appears at numerous levels of system function, from the cellular to the cognitive.

Thayer and Lane (2000) also take a dynamic systems perspective on emotion and behavioral subsystems. In the service of goal-directed behavior and in the context of a behavioral system, they see these organized into coordinated assemblages that can be described by a small number of control parameters. This is much like the factors of factor analysis, which reveal the latent structure among a set of questionnaire items thereby reducing or mapping the higher dimensional item space into a lower dimensional factor space. In their view, emotions may represent preferred configurations in a larger 'state-space' of a possible behavioral repertoire of the organism. From their perspective, disorders of affect represent a condition in which the individual is unable to select the appropriate response, or to inhibit the inappropriate response, so that the response selection mechanism is somehow corrupted.

Gilbert (2001) suggests that a canonical form of such 'corruption' is the excitation of modes that, in other circumstances, represent 'normal' evolutionary adaptations, a matter to which we will return.

Panskepp (2003) has argued that emotion represents a primary form of consciousness, based in early-evolved brain structures, which has become convoluted with what we here describe as a later-developed global neuronal workspace. To anticipate the argument somewhat, we are going to suggest that the convolution with GNW consciousness involves quite a large number of other cognitive biological and social submodules as well.

2.6   Sociocultural network

Humans are particularly noted for a hypersociality that inevitably enmeshes us all in group processes of decision and a collective cognitive behavior within a social network, tinged by an embedding shared culture. For humans, culture is truly fundamental. Durham (1991) argues that genes and culture are two distinct but interacting systems of inheritance within human populations. Information of both kinds has influence, actual or potential, over behaviors, which creates a real and unambiguous symmetry between genes and phenotypes on the one hand, and culture and phenotypes on the other. Genes and culture are best represented as two parallel lines or tracks of hereditary influence on phenotypes.

Much of hominid evolution can be characterized as an interweaving of genetic and cultural systems. Genes came to encode for increasing hypersociality, learning, and language skills. The most successful populations displayed increasingly complex structures that better aided in buffering the local environment (e.g. Bonner, 1980).

Successful human populations seem to have a core of tool usage, sophisticated language, oral tradition, mythology, music, and decision making skills focused on relatively small family/extended family social network groupings.

More complex social structures are built on the periphery of this basic object (e.g. Richerson and Boyd, 1995, 2004). The human species' very identity may rest on its unique evolved capacities for social mediation and cultural transmission. These are particularly expressed through the cognitive decision making of small groups facing changing patterns of threat and opportunity, processes in which we are all embedded and all participate.

This listing, in concert with our general focus on consciousness, suggests a more comprehensive picture of chronic mental and physical disorder than is current.

2.7   Comorbidity

Chapter 3 discussed briefly the problem of dividing the full set of possible responses of a cognitive process into resting and active sets, respectively labeled B_0 and B_1. Like the 'second order selection' we proposed for the dynamic threshold of consciousness, it seems likely that a higher order cognitive module must act to define which states are to be labeled as resting and active. This is because, depending on the patterns of threat or opportunity facing the organism, different 'languages of thought' are appropriate at different times. Perceived threat, for example, requires activation of the HPA axis as 'normal' for the duration of that threat. This suggests existence of, in addition to consciousness as a higher order function, a 'zero mode identification' cognitive module (ZMI), whose dysfunction through improper identification of a typically excited mode as a resting mode, can result in chronic disease. See R. Wallace (2003, 2004) for an extended discussion.

The idea is, basically, a generalization of Gilbert's (2001) mechanism for emotional disorder, i.e. having some 'normal' evolutionary adaptation become pathologically persistent or inappropriately activated. An example might be long-lasting emergency hypervigilance in anxiety disorder. We suppose most cognitive modules subject to similar problems, likely with mutually-reinforcing 'crosstalk' between them: comorbidity.

If Y represents the information source dual to ZMI in generalized cognition, and if Z is the information source characterizing 'structured psychosocial stress', an embedding context, the mutual information between them $I(Y;Z) = H(Y) - H(Y|Z)$, serves as a splitting criterion for pairs of linked paths of states.

Structured psychosocial stress is usually long-term, continually affecting individuals, families, and communities at all stages of life. Certain physiologically excited modes are thus likely to be continually activated during the life course, from gestation, birth, and growth, through senescence. A typical example would be growing up or living under an Apartheid system, a Manichean world, divided into 'good' and 'bad', 'black' and 'white' (Fanon, 1966; Massey and Denton, 1998; Memmi, 1969). Other examples would include American-style 'regulated poverty', the British 'class' system, nonunionized workplaces, prisons, the military during combat, and so on.

Parametize the coupling between these interacting information sources, writing $I(Y;Z) = I[K]$, with structured psychosocial stress as the embedding context.

Invocation of the mathematical homology of Section 2.7 permits imposition of renormalization formalism resulting in punctuated phase transition depending on K.

Socioculturally constructed and structured psychosocial stress, in this model having both 'grammar' and 'syntax', can be viewed as entraining the function of zero mode identification when the coupling with stress exceeds a threshold. More than one threshold appears likely, accounting, perhaps, for the often staged nature of 'environmentally caused' disorders. These should result in a series of collective, but highly systematic, 'tuning failures' in the Rate Distortion sense, that represent a literal image of the structure of imposed psychosocial stress written upon the ability of the ZMI system to characterize a 'normal' mode of excitation. This causes a mixed atypical and usually transient state to become permanent, producing comorbid mental and chronic physical disorder. The process may have both cross-sectional and longitudinal structure, with the latter accounting for 'critical periods' in the onset of developmental disorders.

Coronary heart disease (CHD) is already understood as a disease of development which begins *in utero*. Work by Barker and colleagues (Barker, 2000; Barker et al., 2002; Eriksson et al., 2000; Godfrey and Barker, 2001; Osmond and Barker, 2000) suggests that those who develop CHD grow differently from others, both in utero and during childhood. Slow growth during fetal life and infancy is followed by accelerated weight gain in childhood, setting a life history trajectory for CHD, type II diabetes, hypertension, and, of course, obesity. Barker (2002) concludes that slow fetal growth might also heighten the body's stress responses and increase vulnerability to poor living conditions later in life. Thus, in his view, CHD is a developmental disorder that originates through two widespread biological phenomena, developmental plasticity and compensatory growth, a conclusion consistent with the work of Smith et al. (1998), who found that deprivation in childhood influences risk of mortality from CHD in adulthood, although an additive influence of adult circumstances is seen in such cases.

Much of the CHD work particularly implicates certain kinds of hypertension as a developmental disorder. As Eriksson et al. (2000) put the matter,

> The association between low birth weight and raised blood pressure in later life has now been reported in more than 50 published studies of men, women, and children. It has been shown to result from retarded fetal growth rather than premature birth. The 'fetal origins' hypothesis proposes that the association reflects permanent resetting of blood pressure by undernutrition in utero.

Asthma fits within a similar perspective. Wright et al. (1998) describe how prospective epidemiological studies show the newborn period is dominated by Th2 reactivity in response to allergens. It is also evident that Th1 memory cells selectively develop shortly after birth, and persist into adulthood in non-atopic subjects. For most children who become allergic or asthmatic, the polarization of their immune systems into an atopic phenotype probably occurs during early childhood. It has been speculated that stress triggers hormones in the early months of life which may influence Th2 cell predominance, perhaps through a direct influence of stress hormones on the production of cytokines that are thought to modulate the direction of immune cell differentiation. There is evidence that parental reports of life stress are associated with subsequent onset of wheezing in children between birth and one year.

Recent work by Collins et al. (2004) explores how the experience of racism can express itself as very low birthweight among African-Americans. He concludes that

> Our study adds to the small but growing evidence of a relation between African American women's exposure to interpersonal racial discrimination and pregnancy outcomes. We found that African American mothers who delivered [very low birthweight] preterm infants were more likely to report discrimination during their lifetime than African American mothers who delivered [normal birthweight] infants at term...the reported lifelong accumulated experiences of interpersonal racial discrimination by African American women constitute an independent risk factor for infant [very low birthweight].

Work by Hirsch (2003) suggests that obesity, which is also seriously epidemic in the USA, is a developmental disorder with roots in utero or early childhood. Hirsch and others have developed a 'set point' or homeostatic theory of body weight, finding that it is the process which determines that 'set point' which needs examination, rather than the homeostasis itself, which is now fairly well understood. Hirsch concludes that the truly relevant question is not why obese people fail treatment, it is how their level of fat storage became elevated, a matter, he concludes, is probably rooted in infancy and childhood, when strong genetic determinants are shaping a still-plastic organism. In this regard, Bjorntorp (2001) finds that embedding psychosocial stress is a principal determinant of obesity at both the individual and population levels.

Somewhat less conclusively, a lively debate rages regarding various possible subforms of psychopathy, a mental disorder characterized by a long history of manipulative, impulsive, and callous antisocial 'cheating' behavior. Mealey (1995) places the disorder in an evolutionary perspective as either a genetically determined or an acquired 'life history strategy' very similar to Nunney's (1999) analysis of cancer, albeit at the social rather than cellular level of interaction. Paris (1993) attempts to provide a comprehensive, integrative, biopsychosocial perturbed 'condition-development' model for personality disorders. Lalumiere et al. (2001), by contrast, find evidence for a strict life-history strategy model, concluding, as a result of studies on children and adolescents, that "If psychopathy is a result of condition-development, the environmental triggers are likely to operate very early". The review by Herpertz et al. (2000) examines the hypothesis that pathologically neglectful parenting and early social rejection contribute to onset of the disorder, particularly in the context of 'individualistic' social structures (Cooke, 1996). We speculate that it is possible to place the 'social cheating' of psychopathy in the same context as Nunney's cellular cheating for cancer, consequently being subject to the standard pattern of gene-environment 'norms of reaction' emerging as structured psychosocial stress acts over the course of child development, probably beginning in utero.

It almost goes without saying that the diagnosis of psychopathy (like other 'personality disorders') is very much concentrated in prison subpopulations. These always have marked ethnic and 'racial' structure as a consequence of formal patterns of discrimination, economic deprivation, and various forms of Apartheid – all constituting structured psychosocial stressors which write literal images of themselves upon their victims either through induction of developmental disorders or as adaptation pressures.

Anxiety disorders have a long history of attribution to developmental factors and early childhood exposures (Bandelow et al., 2002). More generally, Egle et al. (2002) find evidence that early biological and psychosocial stress in childhood is associated with long-term vulnerability to various mental and physical diseases. Research findings have, in their view, accumulated on those emotional, behavioral and psychobiological factors responsible for the mediation of lifelong consequences including increased risk of somatization and other mental disorders such as anxiety, depression and personality disorders. These often result in high-risk behaviors that are associated with physical disease – cardiovascular disorders, stroke, hepatitis C, type 2 diabetes, chronic lung disease, as well as with aggressive behavior.

These case histories appear to present specific instances of a comprehensive general phenomenon affecting the etiology of the larger spectrum of chronic and comorbid mental and physical disorders, in the sense that structured psychosocial stress can literally write an image of itself upon the developing child, and if acute enough, on the adult, initiating trajectories to comorbid mental and chronic physical disorder.

Such disorder must inevitably constitute a powerful context for individual consciousness: mind-body dysfunction will always have profound impact on conscious experience and the possibilities for voluntary action.

The next example suggests that such interactions may involve 'second order' as well as direct effects.

2.8   Schizophrenia

Schizophrenia appears to fall broadly within the paradigm of a developmental cognitive disorder (Lewis and Levitt, 2002; Allin and Murray, 2002). Within the United Kingdom schizophrenia is, however, significantly more prevalent among Afro-Caribbean immigrants subject to chronic unemployment, early separation from parents, and perhaps racial discrimination, when compared with non-migrants of either majority or minority ethnicity (Mallett et al., 2002). For the U.S. there is some controversy as to the propensity of majority clinicians to over-diagnose schizophrenia among minority patients, perhaps masking underlying demographic patterns. As Gaughran et al. (2002) note, however, there is good evidence of immune activation in schizophrenia. Up to a third of patients has an autoimmune condition clinically unrelated to their psychiatric illness, and first degree relatives of people with schizophrenia also have increased incidence of autoimmune disease.

Torrey and Yolken (2001) note the similarities and contrasts between schizophrenia and rheumatoid arthritis. Both are chronic, persistent diseases displaying lifelong prevalence and a relapsing and remitting course. Both are felt to involve environmental insults occurring in genetically susceptible individuals, and their diagnosis depends upon syndromal diagnostic criteria which have been developed by committees and have changed over time. Many studies, however, have observed a striking inverse correlation – an 'anticomorbidity' – between the two diseases, although both are believed to run in families, with a population prevalence of about one percent. That is, people with schizophrenia seem less likely to suffer from rheumatoid arthritis, although perhaps more likely to suffer autoimmune disease in general.

This begins to resemble the retinal 'nonorthogonal eigenmode' patterns discussed above.

Grossman et al. (2003) describe how the recent emphasis on schizophrenia as a developmental disorder has focused on characterizing the role of non-genetic factors in the development of symptom patterns. Certain prenatal and perinatal environmental exposures, including maternal stress and malnourishment, and obstetric complications such as low birth weight, have been reported to be associated with increased susceptibility to the disorder. Increased incidence has also been reported in children born to mothers who experienced infection from influenza or rubella during the second trimester of pregnancy. Thus early neurodevelopmental processes may be compromised, laying groundwork for disorder when taxed by later developmental demands, for example those associated with the stressful periods of social development in childhood and adolescence.

Rothermundt et al. (2001) further summarize at some length the case for both the infection and autoimmune hypotheses regarding onset of schizophrenia.

Torrey and Yolken (2001) conclude that the negative association between schizophrenia and rheumatoid arthritis may depend on the timing of some critical exposure, e.g. that exposure in utero or childhood produces schizophrenia, while exposure in adulthood produces rheumatoid arthritis. A slightly different hypothesis, consistent with the mathematical exercises above, is that rheumatoid arthritis and schizophrenia characterize different atypical mixed eigenmodes falsely and recurrently identified as zero states by the progressive failure of the ZMI module. Such would tend to be mutually exclusive, although not absolutely so since the eigenmodes are not orthogonal.

A broadly similar pattern has been commented on by Karlsson et al. (2001), who found homologous sequences of the HERV-W family of endogenous retroviruses in the cerebro-spinal fluid of newly-diagnosed individuals with schizophrenia and in other subjects having multiple sclerosis. Karlsson et al. (2001) speculate it is possible that individuals with schizophrenia and multiple sclerosis undergo the activation of similar retroviral sequences but differ in terms of genetically determined responses to the retroviral activation. Schizophrenia and multiple sclerosis are distinct clinical entities and have different pathological manifestations, gender ratios, and clinical courses, but share a number of epidemiological features including age of onset, seasons of birth, and geographic distributions. In addition, however, some patients display clinical manifestations of both diseases.

Similarly, rigorous studies by Dupont et al. (1986), Gulbinat et al. (1992) and Mortinsen (1989, 1994) followed large Danish and Dutch cohorts of patients with schizophrenia. When adjusted for smoking patterns, these showed marked and highly significant reduction in a broad variety of cancers. More recent work by Cohen et al. (2002) adjusted for age, race, gender, marital status, education, net worth, smoking, and hospitalization in the year before death, for a large US sample likewise found greatly reduced risk of cancer among persons diagnosed with schizophrenia. Catts and Catts (2000) speculate that such results are driven by hyperactivation of the p53 tumor suppressor/apoptosis gene during neurodevelopment, causing long-term developmental dysfunction, while Teunis et al. (2002) suggest, from animal model studies, that the hyperreactive dopaminergic system characteristic of schizophrenia inhibits tumor vascularization.

These examples, again, strongly suggest a 'nonorthogonal eigenmode' pattern in which the ZMI module, including both immune function and the larger system of tumor control mechanisms within a unified and broadly cognitive structure, fails in a systematic manner, producing characteristic spectra of co- and antico- morbidities among different dysfunctions.

It appears that, at the population level, structured psychosocial stress can also exert a 'higher order effect,' producing different spectra of co- and anticomorbidities between schizophrenia and other disorders within powerful and marginalized subgroups. This prediction, which extends the analysis of Section 2.7, should be empirically testable.

3. The Hierarchical Cognitive Model

An essential characteristic of cognition in this formalism involves a function $h(x)$ which maps a (convolutional) path $x = a_0, a_1, ..., a_n, ...$ onto a member of one of two disjoint sets, B_0 or B_1. Thus respectively, either (1) $h(x) \in B_0$, implying no action taken, or (2), $h(x) \in B_1$, and some particular response is chosen from a large repertoire of possible responses. The problem of defining these two disjoint sets arises, and a 'higher order cognitive module' seems needed to identify what constituted B_0 (the set of 'normal' states), a matter which may vary according to the challenges faced by the organism.

We suppose that higher order cognitive module, Zero Mode Identification, interacts with an embedding language of structured psychosocial stress (or other systemic perturbation) and, instantiating a Rate Distortion image of that embedding stress, begins to include one or more members of the set B_1 into the set B_0. Recurrent 'hits' on that aberrant state would be experienced as episodes of highly structured comorbid mind/body pathology.

Empirical tests of this hypothesis, however, all quickly lead into real-world regression models or their time series variants, involving the interrelations of measurable biomarkers, beliefs, behaviors, neural correlates, reported thoughts and feelings, and so on. This has certain theoretical as well as practical consequences, and a recapitulation of regression is in order, much in the spirit as was consideration of the visual retina.

The General Linear Model so familiar to empirical researchers is based on several critical assumptions. In the simplest case, following Snedecor and Cochran (1979, p. 141), there are three essential restrictions on the relation between independent and dependent variates X and Y:

1. For each selected X there is a Normal distribution of Y from which the sample value of Y is drawn at random. If desired, more than one Y may be drawn from each distribution.

2. The population of values of Y corresponding to a selected X has a mean μ that lies on the straight line

$$\mu = \alpha + \beta(X - \hat{X}) = \alpha + \beta x,$$

where α and β are parameters to be estimated. $\hat{X}$ is the mean of X.

3. In each population the standard deviation of Y about its mean $\alpha + \beta x$ has the same value, assumed constant as x varies.

The mathematical model is specified concisely by the equation

$$Y = \alpha + \beta x + \epsilon.$$

where ϵ is a random variable drawn from an appropriate Normal distribution.

The central problem then becomes the statistical estimation of the parameters α and β from observational data.

Variants of this model range from multiple regression, to canonical correlation, and, more recently, our own work on estimating system response to external perturbation (D. Wallace and R. Wallace, 2000). Similar methods can, of course, be used for more complicated 'linearizable' problems, for example fitting to polynomials or exponentials in x.

A particularly important generalization of linear regression is the hierarchical linear model in which the parameters of a set of regressions conducted at one scale are treated as variables in a regression conducted at a larger scale. An example would be models relating health indices to income measures for individuals which are each conducted at the Zip Code level across a set of Zip Codes differing in some set of characteristic measures. Creating a hierarchical model would involve a regression expressing the slope and intercept 'constants' of the individual-scale regressions as a function of the Zip Code characteristics. Again, Byrk and Raudenbusch (2001) is the classic reference.

Indeed, as Anderson (1971) comments, many of the statistical techniques used in time series analysis are actually those of regression analysis – classical least squares theory – or adaptations or analogs of them, often translated from time-domain to frequency domain via Fourier or related transforms.

All such methods are, however, organized around the Central Limit Theorem.

Languages, as information sources, are different, being much more highly structured, and cannot be addressed in quite the same manner. Here we, in effect, propose a General Cognitive Model for punctuation and other behavior in cognitive systems based, not on the Central Limit Theorem, but rather on the Shannon-McMillan Theorem, as modulated by the obvious homology with free energy density. The trick is to associate a cognitive process with a dual information source which is adiabatically piecewise memoryless ergodic, using renormalization formalism at punctuation, and generalized Onsager relations away from punctuation. The model may, itself, be iterated to higher order in renormalization parameters, much as the HLM generalizes the GLM.

The central problem of the General Cognitive Model or its hierarchical extension, the HCM, then becomes, in analogy with the GLM and the HLM, the estimation, from observational data, of the renormalization relation and its 'universality class' parameters, which may be both tunable and distributed, and, away from punctuation, the generalized Onsager relations. This must be done in the context of possible complications resulting from second-order punctuation which is the precise analog of the HLM.

The different possible renormalization schemes or Onsager relations for the GCM or the HCM are analogous to different possible polynomial or exponential fittings in the GLM or its hierarchical extension.

Generalized parameter estimation for such models appears fiendishly difficult, except perhaps under very restricted experimental conditions.

In defense of the proposed empirical technique, cognitive and conscious processes are themselves fiendishly complicated, and what we have done may well be as simple as things can realistically be made – the cognitive equivalent of a straight line regression relation or simple time series analysis.

As is often true for the GLM, analysis of 'residuals' from fitting a GCM or HCM might well provide critical scientific insight: the GCM could serve as a compelling theoretical benchmark against which to compare real data. However, the realities of experimental technique necessarily require a strong interrelation between the proposed GCM and the standard GLM.

4.   Evading the mereological fallacy

We have constructed a punctuated, information-dynamic statistical model of the global neuronal workspace – the HCM – incorporating a second-order and similarly punctuated universality class tuning linked to detection and interpretation of structured external signals. The model, which features a 'tunable retina' atlas/manifold topology, suggests that tuning the punctuated activation of attention to those signals permits more rapid and appropriate response, but at increased physiological or other opportunity cost: unconscious processing is clearly more efficient, if the organism can get away with it. On the other hand, if the environment is threatening, the organism cannot always get away with it, suggesting a strong evolutionary imperative for a dynamic global neural workspace.

Linkage across individual dynamic workspaces – human hypersociality in the context of an embedding epigenetic system of cultural inheritance – would be even more adaptationally efficient. Indeed, equations (5.9-5.11) suggest the possibility of very strong linkage of individual consciousness and physiology to embedding sociocultural network phenomena, ultimately producing an extended model of consciousness which does not fall victim to the mereological fallacy.

In just this regard Nisbett et al. (2001), following in a long line of research (Markus and Kitayama, 1991, and the summary by Heine, 2001), review an extensive literature on empirical studies of basic cognitive differences between individuals raised in what they call 'East Asian' and 'Western' cultural heritages, which they characterize, respectively, as 'holistic' and 'analytic.' They find:

1. Social organization directs attention to some aspects of the perceptual field at the expense of others.

2. What is attended to influences metaphysics.

3. Metaphysics guides tacit epistemology, that is, beliefs about the nature of the world and causality.

4. Epistemology dictates the development and application of some cognitive processes at the expense of others.

5. Social organization can directly affect the plausibility of metaphysical assumptions, such as whether causality should be regarded as residing in the field vs. in the object.

6. Social organization and social practice can directly influence the development and use of cognitive processes such as dialectical vs. logical ones.

Nisbett et al. (2001) conclude that tools of thought embody a culture's intellectual history, that tools have theories built into them, and that users accept these theories, albeit unknowingly, when they use these tools.

Individual consciousness – exemplified by the dynamic global neuronal workspace model – appears to be profoundly affected by cultural, and perhaps developmental, context, and, we aver, by patterns of embedding psychosocial stress. These are all matters subject to a direct empirical study which may lead to an extension of the concept particularly useful in understanding certain forms of psychopathology.

Current dynamic systems models of neural networks, or their computer simulations, simply do not reflect the imperatives of Adams' (2003) informational turn in philosophy. Dynamic systems models based on differential equations, or their difference equation realizations on computers, are pursued nonetheless because they have a history of intense and continuous intellectual development going back to Isaac Newton. Hence very little new mathematics needs to be done, and one can look up most required results in the textbooks, which are quite sophisticated by now. By contrast, rigorous probability theory is perhaps a hundred years old, its information theory subset has seen barely a half century, and the tunable retina atlas/manifold formalism is still under development. Consequently the mathematics cannot always be looked up, and must often be created de novo, with considerable difficulty. Relentless application of the dynamic systems paradigm to consciousness reminds one, not originally, of a drunk looking for his lost car keys under a street lamp 'because the light is better here'.

Nisbett's caution that tools of thought embody a cultural history whose built-in theories users implicitly adopt is no small matter: dynamical systems theory carries with it more than just a whiff of the 18th Century mechanical clock, while statistical mechanics models of neural networks cannot provide natural linkage with the sociocultural contexts which carry the all-important human epigenetic system of heritage (Richerson and Boyd, 2004).

Again, to paraphrase Heine (2001), the extreme nature of U.S. cultural individualism raises the specter that work based on late 20th Century American research not only stands the risk of developing an understanding of consciousness that is particular to that culture, but also of developing an understanding of consciousness that is peculiar in the context of the world's cultures.

Even when going somewhat beyond dynamic systems theory, most current applications of information theory to the global neuronal workspace appear to have strayed far indeed from the draconian structural discipline imposed by the asymptotic limit theorems of the subject. Information measures are of relatively little interest in and of themselves, serving primarily as grist for the mills of splitting criteria between high and low probability sets of behavioral paths. This is the central mechanism whose extension, using a homology with free energy density, permits exploration of tunable punctuation in a manner consistent with the program described by Adams (2003).

According to the mathematical ecologist E.C. Pielou (1976, p.106), the legitimate purpose of mathematical models is to raise questions for empirical study, not to answer them, or, as one wag put it, "all models are wrong, but some models are useful." The natural emergence of tunable punctuation in our treatment, albeit at the expense of elaborate renormalization calculations at transition, and generalized Onsager relations away from it, suggests the possible utility of the theory in future empirical studies of consciousness. The car keys really may have been lost in the dark parking lot down the street, but here is a new flashlight.

We have outlined an empirically-testable approach to modeling consciousness which returns with a resounding thump to the classic asymptotic limit theorems of communication theory, and suggests further the necessity of incorporating the effects of embedding structures of psychosocial stress and culture. The theory suffers from a painful grandiosity, claiming to incorporate matters of cognition, consciousness, social system, psychopathology, and culture into a single all-encompassing model. To quote from a recent review of Bennett and Hacker's new book, (Patterson, 2003), however, contemporary neuroscience itself may suffer a more pernicious and deadly form of that disorder for which our approach is, in fact, the antidote:

> [Bennett and Hacker] argue that for some neuroscientists, the brain does all manner of things: it believes (Crick); interprets (Edelman); knows (Blakemore); poses questions to itself (Young); makes decisions (Damasio); contains symbols (Gregory) and represents information (Marr). Implicit in these assertions is a philosophical mistake, insofar as it unreasonably inflates the conception of the 'brain' by assigning to it powers and activities that are normally reserved for sentient beings... these claims are not false; rather they are devoid of sense.

This is but one example of a swelling critical chorus which will grow markedly in virulence and influence. Our development, or some related version, leads toward explicit incorporation of the full 'sentient being' into observational studies of consciousness. For humans, whose hypersociality is both glory and bane, this particularly involves understanding the effects of the embedding social and cultural system of epigenetic inheritance on immediate conscious experience – searching for the torus and the sphere.

The bottom line would seem to be, very much in the spirit of Pielou's caution regarding the utility of mathematical models, the urgent necessity of extending the perspectives of Markus and Kitayama (1991) and Nisbett et al. (2001) to brain imaging and other empirical studies of consciousness and its disorders, and expanding the global neuronal workspace model accordingly, a matter which our development here suggests is indeed possible, if not straightforward.

The evolutionary anthropologist Robert Boyd has commented that "Culture is as much a part of human biology as the enamel on our teeth." An appropriate paraphrase might well read "Culture is as much a part of human consciousness as the neurons in our brains."

Although we have left the intellectual history of these matters to Barrs and others, another way of putting it is to say that William James' stream of consciousness has cultural riverbanks and historical shoals which define its speed, depth, and possible directions.

The scientific mapping of that cultural hydrogeography is overdue.

Chapter 7

REFERENCES

Adams F., (2003), The informational turn in philosophy, *Minds and Machines*, 13:471-501.

Ademi C., C. Ofria, and T. Collier, (2000), Evolution of biological complexity, *Proceedings of the National Academy of Sciences*, 97:4463-4468.

Albert R. and A. Barabasi, (2002). Statistical mechanics of complex networks. Rev. Mod. Phys. 74:47-97.

Allin M., and R. Murray, (2002), Schizophrenia: a neurodevelopmental or neurodegenerative disorder?, *Current Opinion in Psychiatry*, 15:9-15.

Anderson T., (1971), *The Statistical Analysis of Time Series*, Wiley, New York.

Aof M. and R. Brown, (1992), The holonomy groupoid of a locally topological groupoid. Topology Appl. 47:97-113.

Arnold V., (1989), *Mathematical Methods of Classical Mechanics*, Springer-Verlag, New York.

Ash R., (1990), *Information Theory*, Dover Publications, New York.

Atick J. and A. Redlich, (1990), Towards a theory of early visual processing, *Neural Computation*, 2:308-320.

Atick J., (1992), Could information theory provide an ecological theory of sensory processing? *Network Computation in Neural Systems*, 3:213-251.

Atlan H. and I. R. Cohen, (1998), Immune information, self-organization and meaning, International Immunology 10:711-717.

Attneave F., (1954), Some informational aspects of visual perception, *Psychological Review*, 61:183-193.

Baars B., (1983), Conscious contents provide the nervous system with coherent, global information. In *Consciousness and Self-Regulation*(Vol. 3)(Davidson R.J. et al. eds.), Plenum Press.

Baars B., (1988)*A Cognitive Theory of Consciousness*, Cambridge University Press.

Baars B., (2002), The conscious access hypothesis: origins and recent evidence, *Trends in Cognitive Sciences*, 6:47-52.

Baars B., and S. Franklin, (2003), How conscious experience and working memory interact, *Trends in Cognitive Science*, doi:10.1016/S1364-6613(03)00056-1.

Balasubramanian V., D. Kimber and M. Berry, (2001), Metabolically efficient information processing, *Neural Computation*, 13:799-815.

Bandelow B., C. Spath, G. Tichauer, A. Broocks, G. Hajak, and E. Ruther, (2002), Early traumatic life events, parental attitudes, family history, and birth risk factors in patients with panic disorder, *Comprehensive Psychiatry*, 43:269-278.

Barker D., (2002), Fetal programming of coronary heart disease, *Trends in Endocrinology and Metabolism*, 13:364-372.

Barker D., T. Forsen, A Uutela, C. Osmond, and J. Erikson, (2002), Size at birth and resilience to effects of poor living conditions in adult life: longitudinal study, *British Medical Journal*, 323:1261-1262.

Barlow H., (1959), Sensory mechanisms, the reduction of redundancy and intelligence, in D. Blake and A. Utley (eds.), *Proceedings of the Symposium on the Mechanization of Thought Processes, Vol. 2*, MIT Press, Cambridge, MA.

Barlow H., (1961), Possible principles underlying the transformation of sensory messages, in *Sensory Communication*, W. Rosenblith ed., MIT Press, Cambridge, MA.

Bar-Yam, Y., (1997), *Dynamics of Complex Systems*, Addison-Wesley, Reading, MA.

Beck C. and F. Schlogl, (1995), *Thermodynamics of Chaotic Systems*, Cambridge University Press.

Behrman E., J. Niemel, J. Stec, and S. Skinner, (1996), A quantum dot neural network, *Proceedings of the Workshop on Physics of Computation*, New England Complex Systems Institute, Cambridge, MA, pp. 22-24.

Benedict R., (1934), *Patterns of Culture*, Houghton Mifflin, New York.

Bennett C., (1988), Logical depth and physical complexity, in *The Universal Turing Machine: A Half-Century Survey*, R. Herkin, ed., pp. 227-257. Oxford University Press.

Bennett M., and P. Hacker, (2003), *Philosophical Foundations of Neuroscience*, Blackwell Publishing.

Binney J., N. Dowrick, A. Fisher, and M. Newman, (1986), The theory of critical phenomena. Clarendon Press, Oxford UK.

Bjelakivic I., T. Kruger R. Siegmund-Schultze, A. Szkola, (2003), Chained typical subspaces – a quantum version of Brieman's theorem, ArXiv, quant-ph/0301177.

Bjelakovic I., T. Kruger, R. Siegmund-Schultze, A. Szkola, (2004), The Shannon-McMillan theorem for ergodic quantum lattice systems, *Inventiones mathematicae*, 155:203-222.

Bjorntorp P., (2001), Do stress reactions cause abdominal obesity and co-morbidities? *Obesity Reviews*, 2:73-86.

Bonner J, (1980), *The Evolution of Culture in Animals*, Princeton University Press, Princeton NJ.

Brenner N., W. Bialek, and R. De Ruyter van Stevenick, (2000), Adaptive rescaling maximizes information transmission, *Neuron*, 26:695-702.

Brown R., (1987), From groups to groupoids, Bull. London Math. Soc. 19:113-134.

Byrk A., and S. Raudenbusch, (2001), *Hierarchical Linear Models: Applications and Data Analysis Methods*, Sage Publications, New York.

Catts V., and S. Catts, (2000), Apoptosis and schizophrenia: is the tumor suppressor gene p53 a candidate susceptibility gene? *Schizophrenia Research*, 41:405-415.

Cohen I.R., (1992), The cognitive principle challenges clonal selection. Immunology Today 13:441-444.

Cohen I.R., (2000), *Tending Adam's Garden: evolving the cognitive immune self*, Academic Press, New York.

Cohen M., B. Dembling, and J. Schorling, (2002), The association between schizophrenia and cancer: a population-based mortality study, *Schizophrenia Research*, 57:139-146.

Collins J., R. David, A. Handler, S. Wall, and S. Andes, (2004), Very low birthweight in African American infants: the role of maternal exposure to interpersonal racial discrimination, *American Journal of Public Health*, 94:2132-2138.

Cooke D., (1996), Psychopathic personality in different cultures: what do we know?, *Journal of Personality Disorders*, 10:23-40.

Cover T. and J. Thomas, (1991), *Elements of Information Theory*. Wiley, New York.

Damasio A., (1989), Time-locked multiregional retroactivation: a systems-level proposal for the neural substrates of recall and recognition, *Cognition*, 33:25-62.

Damasio A., (1998), Emotion in the perspective of an integrated nervous system, *Brain Research Reviews*, 26:83-86.

Dehaene S. and L. Naccache, (2001), Towards a cognitive neuroscience of consciousness: basic evidence and a workspace framework, Cognition 79:1-37.

Dembo A., and O. Zeitouni, (1998), *Large Deviations: Techniques and Applications, 2nd. Ed.*, Springer-Verlag, New York.

Dimitrov A. and J.Miller, (2001), Neural coding and decoding: communication channels and quantization. Network: Comput. Neural Syst. 12:441-472.

Dretske F., (1981), *Knowledge and the flow of information*, MIT Press, Cambridge, MA.

Dretske F., (1988), *Explaining behavior*, MIT Press, Cambridge, MA.

Dretske F., (1992), What isn't wrong with folk psychology, *Metaphilosophy*, 29:1-13.

Dretske F., (1993), Mental events as structuring causes of behavior. In *Mental causation*, (ed. A. Mele and J. Heil), pp. 121-136, Oxford University Press.

Dretske F., (1994), The explanatory role of information, *Philosophical Transactions of the Royal Society A*, 349:59-70.

Dupont A., O. Jensen, E. Stromgren, and A. Jablonsky, (1986), Incidence of cancer in patients diagnosed as schizophrenic in Denmark, in: ten Horn, G., R. Giel, W. Gulbinat, and W. Henderson (eds.), *Psychiatric Case Registers in Public Health*, Elsevier, Amsterdam, 229-239.,

Durham W., (1991), *Coevolution: Genes, Culture, and Human Diversity*, Stanford University Press, Palo Alto, CA.

Edelman G., (1989), *The Remembered Present*, Basic Books, New York.

Edelman G. and G. Tononi, (2000), *A Universe of Consciousness*, Basic Books, New York.

Egle U., J. Hardt, R. Nickel, B. Kappis, and S. Hoffmann, (2002), Long-term effects of adverse childhood experiences – actual evidence and needs for research, *Zeitshrift fur Psychosomatische Medizin und Psychotherapie*, 48:411-434.

Eldredge N., (1985), *Time Frames: The Rethinking of Darwinian Evolution and the Theory of Punctuated Equilibria*, Simon and Schuster, New York.

Elitzur A, (1996), Life's emergence is not an axiom: a reply to Yockey, *Journal of Theoretical Biology*, 180:175-180.

Eriksson J., T. Forsen, J. Tuomilehto, C. Osmons, and D. Barker, (2000), Fetal and childhood growth and hypertension in adult life, *Hypertension*, 39:790-794.

Fanon F., (1966), *The Wretched of the Earth*, Grove Press, Boston.

Feller W., (1970), *An Introduction to Probability Theory and its Applications*, John Wiley and Sons, New York.

Feynman R., and A. Hibbs, (1965), *Quantum Mechanics and Path Integrals*, McGraw-Hill, New York, NY.

Feynman R., (1996), *Feynman Lectures on Computation*, Addison-Wesley, Reading, MA.

Feynman R., (1998), *Statistical Mechanics*, Perseus Books, Reading, MA.

Forlenza M. and A. Baum, (2000), Psychosocial influences on cancer progression: alternative cellular and molecular mechanisms, *Current Opinion in Psychiatry*, 13:639-645.

Fredlin M. and A. Wentzell, (1998), *Random Perturbations of Dynamical Systems*, Springer-Verlag, New York.

Freeman W. (1991), The physiology of perception, *Scientific American*, 264:78-85.

Gaughran F., E. O'Neill, P. Sham, R. Daly, and F. Shanahan, (2002), Soluble Il-2 receptor levels in families of people with schizophrenia, *Schizophrenia Research*, 56:235-239.

Geertz C., (1973), The growth of culture and the evolution of mind. In C. Geertz, *The interpretation of cultures* (pp. 55-87), Basic Books, New York.

Gilbert P., (2001), Evolutionary approaches to psychopathology: the role of natural defenses, *Australian and New Zealand Journal of Psychiatry*, 35:17-27.

Gillooly J., A. Allen, G. West and J. Brown (2004), Metabolic rate calibrates the molecular clock: reconciling molecular and fossil estimates of evolutionary divergence. ArXiv paper q-bio-0404027.

Godfrey K, and D. Barker, (2001), Fetal programming and adult health, *Public Health and Nutrition*, 4:611-624.

Gould, S. and N. Eldredge (1977), Punctuated equilibria: the tempo and mode of evolution reconsidered, *Paleobiology*, 3:115-151.

Grossman A., J. Churchill, B. McKinney, I. Kodish, S. Otte, and W. Greenough, (2003), Experience effects on brain development: possible contributions to psychopathology, *Journal of Child Pschology and Psychiatry*, 44:33-63.

Grossman Z., (1989), The concept of idiotypic network: deficient or premature? In: H. Atlan and I.R. Cohen (eds.), *Theories of Immune Networks*, Springer Verlag, Berlin, p. 3852.

Grossman Z., (1992a), Contextual discrimination of antigens by the immune system: towards a unifying hypothesis, in: A. Perelson and G. Weisbch, (eds.), *Theoretical and Experimental Insights into Immunology*, Springer-Verlag, New York.

Grossman Z., (1992b) *International Journal of Neuroscience*, 64:275.

Grossman Z., (1993) Cellular tolerance as a dynamic state of the adaptable lymphocyte, *Immunology Reviews*, 133:45-73.

Grossman Z., (2000), Round 3, *Seminars in Immunology*, 12:313-318.

Gulbinat W., A. Dupont, A. Jabolinsky, O. Jensen, A. Marsella, Y. Nakane, and N. Sartorius, (1992), Cancer incidence of schizophrenia patients. Results of record linkage studies in three countries. *British Journal of Psychiatry*, 161 Suppl. 18:75-85.

Hameroff S., (2001), Consciousness, the brain, and spacetime geometry, *Annals of the New York Academy of Sciences*, 929:74-104.

Hameroff S., A. Nip, M. Porter, and J. Tuszynskj, (2002), Conduction pathways in microtubules, biological quantum computation, and consciousness, *BioSystems*, 64:149-168.

Hartl D. and A. Clark, (1997). *Principles of Population Genetics*, Sinaur Associates, Sunderland MA.

Heine S., (2001), Self as cultural product: an examination of East Asian and North American selves, *Journal of Personality*, 69:881-906.

Herpertz S. and H. Sass, (2000), Emotional deficiency and psychopathy, *Behavioral Sciences and thye Law*, 18:567-580.

Hiai F., D. Petz, (1994), Entropy densities for algebraic states, *Journal of Functional Analysis*, 125:287-308.

Hirsch J., (2003), Obesity: matter over mind? *Cerebrum*, 5:7-18.

Ingber L., (1982), Statistical mechanics of neocortical interactions: I. Basic formulation *Physica D*, 5:83-107.

Ingber L., (1992), Generic mesoscopic neural networks based on statistical mechanics of neocortical interactions *Physical Review A*, 45:2183-2186.

Jones I. and J. Blackshaw, (2000), An evolutionary approach to psychiatry, *Australian and New Zealand Journal of Psychiatry*, 34:8-13.

Karlsson H., S. Bachmann, J. Schroder, J. McArthur, E. Torrey, and R. Yolken, (2001), Retroviral RNA identified in the cerebrospinal fluids and brains of individuals with schizophrenia, *Proceedings of the National Academy of Sciences*, 98:4634-4639.

Khinchine A., (1957), *The Mathematical Foundations of Information Theory*, Dover Publications, New York.

King C., and A. Lesniewski, (1998), Quanatum sources and a quantum coding theorem, *Journal of Mathematical Physics*, 39:88-101.

Lalumiere M., G. Harris, and M. Rice, (2001), Psycopathy and developmental instability, *Evolution and Human Behavior*, 22:75-92.

Laughlin S., and T. Sejnowski, (2003), Communication in neuronal networks, *Science*, 301:1870-1874.

Lennie P., (2003), The cost of cortical computation, *Current Biology*, 13:493-497.

Lerner Y., M. Harel, and R. Malach, (2004), Rapid completion effects in human high-order visual areas, *Neuroimage*, 21:516-526.

Levins R. and R. Lewontin, (1985). *The Dialectical Biologist*, Harvard University Press, Cambridge MA.

Levins R, (1998), The internal and external in explanatory theories. Science as Culture. 7:557-582.

Levy W., and R. Baxter, (1996), Energy efficient neural codes, *Neural Computation*, 8:531-543.

Lewis D. and P. Levitt, (2002), Schizophrenia as a disorder of neurodevelopment, *Annual Reviews of Neuroscience*, 25:409-432.

Lewontin R., (2000), *The Triple Helix: Gene, Organism and Environment*. Harvard University Press, Cambridge MA.

Llinas R. and U. Ribary, (2001), Consciousness and the brain: the thalamocortical dialogue in health and disease, *Annals of the New York Academy of Science*, 929:166-175.

Lopez-Ruiz R., Y. Moreno, S. Boccaletti, D. Hwang, and A. Pacheco, (2004), Awaking and sleeping in a complex network. ArXiv document nlin.AO/0406053.

Luchinsky, D., (1997), On the nature of large fluctuations in equilibrium systems: observations of an optimal force, *Journal of Physics A Letters*, 30:L577-L583.

Mallett R., J. Leff, D. Bhugara, D. Pang, and J. Zhao, (2002), Social environment, ethnicity, and schizophrenia: a case-control study, *Social Psychiatry and Epidemiology*, 37:329-335.

Markus H., and S. Kitayama, (1991), Culture and the self – implications for cognition, emotion, and motivation, *Psychological Review*, 98:224-253.

Massey, D., and N. Denton, (1998), *American Apartheid*, Harvard University Press, Cambridge, MA.

McCauley L., (1993), *Chaos, Dynamics, and Fractals: An Algorithmic Approach to Deterministic Chaos*, Cambridge University Press, UK.

Mead M., (1975), Review of 'Darwin and facial expression', *Journal of Communication*, 25:209-213.

Mealey L., (1995), The sociobiology of sociopathy: an integrated evolutionary model, *Behavioral and Brain Sciences*, 18:523-599.

Memmi A., (1969), *Dominated Man*, Beacon Press, Boston, MA.

Mortensen P., (1989), The incidence of cancer in schizophrenic patients, *Journal of Epidemiolgy and Community Health*, 43:43-47.

Mortensen P., (1994), The occurrence of cancer in first admitted schizophrenic patients, *Schizophrenia Research*, 12:185-194.

Nisbett R., K. Peng, C. Incheol, A. Norenzayan, (2001), Culture and systems of thought: holistic vs. analytic cognition, Psychological Review, 108:291-310.

Nisbett R., and T. Masuda, (2003), Culture and point of view, *Proceedings of the National Academy of Sciences*, 100:11163-11170.

Norenzayan A., and S. Heine, (2004), Psychological universals across cultures: what are they and how can we know? Submitted.

Nunney, L., (1999), Lineage selection and the evolution of multistage carcinogenesis, *Proceedings of the Royal Society, B*, 266:493-498.

Onsager L. and S. Machlup, (1953), Fluctuations and irreversible processes, *Physical Review*, 91:1505-1512.

Osmond C. and D. Barker, (2000), *Environmental Health Perspectives*, 108, Suppl 3:545-553.

Panskepp J., (2003), At the interface of the affective, behavioral, and cognitive neurosciences: Decoding the emotional feelings of the brain, *Brain and Cognition*, 52:4-14.

Paris J., (1993), Personality disorders: a biopsychosocial model, *Journal of Personality Disorders*, 7:53-99.

Patterson D., (2003), Book review of *Philosophical Foundations of Neuroscience*, *Notre Dame Philosophical Reviews*, 2003.09.10.
http://ndpr.icaap.org/content/archives/2003/9/.

Park H., S. Amari, and K. Fukumizu, (2000), Adaptive natural gradient learning algorithms for various stochastic models, Neural Networks, 13:755-765.

Parker A., T. Gedeon, and A. Dimitrov, (2003), Annealing and the rate distortion problem, in S. Thrun, S. Becker, and K. Obermayer, eds., *Advances in Neural Information Processing Systems 15*, 969-976, MIT Press, Cambridge, MA.

Penrose R., (2001), Consciousness, the brain, and spacetime geometry: an addendum. Some new developments on the Orch OR model for consciousness, *Annals of the New York Academy of Sciences*, 929:105-110.

Pielou E, (1977), *Mathematical Ecology.* John Wiley and Sons, New York.

Podolsky S. and A. Tauber, (1997), *The generation of diversity: Clonal selection theory and the rise of molecular biology*, Harvard University Press, Cambridge, MA.

Rau H. and T. Elbert, (2001), Psychophysiology of arterial baroreceptors and the etiology of hypertension, *Biological Psychology*, 57:179-201.

Repetti R., S. Taylor, and T. Seeman, (2002), Risky families: family social environments and the mental and physical health of offspring, *Psychological Bulletin*, 128:330-366.

Richerson P. and R. Boyd, (1995), The evolution of human hypersociality. Paper for Ringberg Castle Symposium on Ideology, Warfare and Indoctrinability (January, 1995), and for HBES meeting.

Richerson P., and R. Boyd, (2004), *Not by Genes Alone: How Culture Transformed Human Evolution*, Chicago University Press.

Ridley M., (1996), *Evolution*, Second Edition, Blackwell Science, Oxford, UK.

Rojdestvenski I. and M. Cottam, (2000), Mapping of statistical physics to information theory with applications to biological systems, *Journal of Theoretical Biology* 202:43-54.

Rothermundt M., V. Arolt, and T. Bayer, (2001), Review of immunological and immunopathological findings in schizophrenia, *Brain, Behavior, and Immunity*, 15:319-339.

Schumacher B., (1995), Quantum Coding, *Physical Review A*, 51:2738-2747.

Schumacher B., (1996), Sending entanglement through noisy quantum channels, *Physical Review A*, 55:2614-2628.

Schwabe L., and K. Obermayer, (2002), Rapid adaptation and efficient coding, *BioSystems*, 67:239-244.

Sergeant C. and S. Dehaene, (2004), Is consciousness a gradual phenomenon? Evidence for an all-or-none bifurcation during the attentional blink. In press, *Psych. Sci.*
Available at: http://www.unicog.orb/biblio/Author/DEHAENE.html.

Shirkov D. and V. Kovalev, (2001), The Bogoliubov renormalization group and solution symmetry in mathematical physics. *Physics Reports* 352:219-249.

Shulman R.G., F. Hyder, and D. Rothman, (2003), Cerebral metabolism and consciousness, *Comptes Rendus Biologies*, 326:253-273.

Smith G. C. Hart, D. Blane, and D. Hole, (1998), Adverse socioeconomic conditions in childhood and cause-speciofic adult mortality: prospective observational study, *British Medical Journal*, 317:1631-1635.

Smith T. and J. Ruiz, (2002), Psychosocial influences on the development and course of coronary heart disease: current status and implications for research and practice, *Journal of Consulting and Clinical Psychology*, 70:548-568.

Snedecor G. and W. Cochran, (1979), *Statistical Methods*, 6th ed., Iowa State University Press, Ames, Iowa, USA.

Sternberg S., (1964), Lectures on Differential Geometry, Prentice-Hall, NJ.

Steyn-Ross M, D. Steyn-Ross, J. Sleigh, and L. Wilcocks, (2001), Toward a theory of the general-anesthetic-induced phase transition of the cerebral cortex> I. A thermodynamic analogy *Physical Review E*, DOI: 10.1103/PhysRevE.64.011917.

Steyn-Ross M., D. Steyn-Ross, J. Sleigh, and D. Whiting, (2003), Theoretical predictions for spatial covariance of the electroencephalographic signal during the anesthetic-induced phase transition: Increased correlation length and emergence of spatial self-organization, *Physical Review E*, DOI: 10.1103/PhysRevE.68.021902.

Tauber A., (1998), Conceptual shifts in immunology: Comments on the 'two-way paradigm.' In K. Schaffner and T. Starzl (eds.), Paradigm changes in organ transplantation, Theoretical Medicine and Bioethics, 19:457-473.

Tegmark M., (2000), Importance of quantum decoherence in brain processes, *Physical Review E*, 61:4194-4206.

Teunis M., A Kavelaars, E. Voest, J. Bakker, B. Ellenborek, A. Cools, and C Heijnen, (2002), Reduced tumor growth, experimental metastasis formation, and angiogenesis in rats with a hyperreactive dopaminergid system, *FASEB Journal* express article 10.1096/fj.02-014fje.

Thayer J. and R. Lane, (2000), A model of neurovisceral integration in emotion regulation and dysregulation, *Journal of Affective Disorders*, 61:201-216.

Thayer J., and B. Friedman, (2002), Stop that! Inhibition, sensitization, and their neurovisceral concomitants, *Scandinavian Journal of Psychology*, 43:123-130.

Timberlake W., (1994), Behavior systems, associationism, and Pavlovian conditioning, *Psychonomic Bulletin*, Rev. 1, 405-420.

Tishby N., F. Pereira, and W. Bialek, (1999), The information bottleneck method, *Proceedings of the 37th Allerton Conference on Communication, Control, and Computing*.

Tononi G., and G. Edelman, (1998), Consciousness and complexity, *Science*, 282:1846-1851.

Torrey E. and R. Yolken, (2001), The schizophrenia-rheumatoid arthritis connection: infectious, immune, or both? *Brain, Behavior, and Immunity*, 15:401-410.

Toth G., C. Lent, P. Tougaw, Y. Brazhnik, W. Weng, W. Porod, R. Liu, and Y. Huang, (1996), Quantum cellular neural networks, *Superlattices and Microstructures*, 20:473-478.

Wallace R., M. Fullilove,, and A. Flisher, (1996), AIDS, violence and behavioral coding: information theory, risk behavior, and dynamic process on core-group sociogeographic networks, *Social Science and Medicine*, 43:339-352.

Wallace D. and R. Wallace, (2000), Life and death in Upper Manhattan and the Bronx: toward an evolutionary perspective on catastrophic social change, *Environment and Planning A*, 32:1245-1266.

Wallace R., (2000), Language and coherent neural amplification in hierarchical systems: Renormalization and the dual information source of a generalized spatiotemporal stochastic resonance, *International Journal of Bifurcation and Chaos* 10:493-502.

Wallace R., (2002a), Immune cognition and vaccine strategy: pathogenic challenge and ecological resilience, *Open Systems and Information Dynamics* 9:51-83.

Wallace R., (2002b), Adaptation, punctuation and rate distortion: non-cognitive 'learning plateaus' in evolutionary process, *Acta Biotheoretica*, 50:101-116.

Wallace R., (2003), Systemic Lupus erythematosus in African-American women: cognitive physiological modules, autoimmune disease, and structured psychosocial stress, *Advances in Complex Systems*, 6:599-629.

Wallace R., (2004), Comorbidity and anticomorbidity: autocognitive developmental disorders of structured psychosocial stress. *Acta Biotheoretica*, 52:71-93.

Wallace R. and D. Wallace, (2004), Structured psychosocial stress and therapeutic failure, *Journal of Biological Systems*, 12:335-369.

Wallace R. and R.G. Wallace, (1998), Information theory, scaling laws and the thermodynamics of evolution, *Journal of Theoretical Biology* 192:545-559.

Wallace R. and R.G. Wallace, (1999), Organisms, organizations and interactions: an information theory approach to biocultural evolution, *BioSystems* 51:101-119.

Wallace R. and R.G. Wallace, (2002), Immune cognition and vaccine strategy: beyond genomics, *Microbes and Infection* 4:521-527.

Wallace R., R.G. Wallace and D. Wallace, (2003), Toward cultural oncology: the evolutionary information dynamics of cancer, *Open Systems and Information Dynamics* 10:159-181.

Wallace R., D. Wallace and R.G. Wallace, (2004), Biological limits to reduction in rates of coronary heart disease: a punctuated equilibrium approach to immune cognition, chronic inflammation, and pathogenic social hierarchy, *Journal of the National Medical Association* 96:609-619.

Weinstein A., (1996), Groupoids: unifying internal and external symmetry, *Notices of the American Mathematical Association*, 43:744-752.

Welchman A. and J. Harris, (2003), Is neural filling-in necessary to explain the perceptual completion of motion and depth information? *Proceedings of the Royal Society of London, B. Biological Sciences*, 270:83-90.

Wilson K.,(1971), Renormalization group and critical phenomena. I Renormalization group and the Kadanoff scaling picture. *Physical Review B*, 4:3174-3183.

Wright R., M. Rodriguez, and S. Cohen, (1998), Review of psychosocial stress and asthma, *Thorax*, 53:1066-1074.

Zur D., and S. Ullmann, (2003), Filling-in of retinal scotomas, *Vision Research*, 43:971-982.

Appendix A
Coarse-Graining

We use a simplistic mathematical picture of an elementary predator/prey ecosystem for illustration. Let X represent the appropriately scaled number of predators, Y the scaled number of prey, t the time, and ω a parameter defining the interaction of predator and prey. The model assumes that the system's 'keystone' ecological process is direct interaction between predator and prey, so that

$$dX/dt = \omega Y$$

$$dY/dt = -\omega X.$$

Thus the predator populations grows proportionately to the prey population, and the prey declines proportionately to the predator population.

After differentiating the first and using the second equation, we obtain the differential equation

$$d^2 X/dt^2 + \omega^2 X = 0,$$

having the solution

$$X(t) = sin(\omega t); Y(t) = cos(\omega t),$$

with

$$X(t)^2 + Y(t)^2 = sin^2(\omega t) + cos^2(\omega t) \equiv 1.$$

Thus in the two dimensional 'phase space' defined by $X(t)$ and $Y(t)$, the system traces out an endless, circular trajectory in time, representing the out-of-phase sinusoidal oscillations of the predator and prey populations.

Divide the $X - Y$ 'phase space' into two components – the simplest 'coarse graining' – calling the halfplane to the left of the vertical Y-axis A and that to the right B. This system, over units of the period $1/(2\pi\omega)$, traces out a stream of A's and B's having a very precise 'grammar' and 'syntax':

$$ABABABAB...$$

Many other such 'statements' might be conceivable, for example,

$$AAAAA..., BBBBB..., AAABAAAB..., ABAABAAAB...,$$

and so on, but, of the obviously infinite number of possibilities, only one is actually observed, is 'grammatical': $ABABABAB....$

Note that finer coarsegrainings are possible within a system, for example dividing phase space in this simple model into quadrants, producing a single 'gramatical' statement of the form $ABCDABCDABCD....$

The obvious, and difficult, question is which coarsegraining will capture the essential behaviors of interest without too much distracting high-frequency 'noise'.

More complex dynamical system models, incorporating diffusional drift around deterministic solutions, or even very elaborate systems of complicated stochastic differential equations, having various 'domains of attraction', i.e. different sets of grammars, can be described by analogous 'symbolic dynamics' (e.g. Beck and Schlogl, 1993, Ch. 3).